Matias Peluffo

Adopting New Zealand Dairy Farm Principles in Argentina

Matias Peluffo

Adopting New Zealand Dairy Farm Principles in Argentina

An Empirical Research on Dairy Systems

VDM Verlag Dr. Müller

Imprint

Bibliographic information by the German National Library: The German National Library lists this publication at the German National Bibliography; detailed bibliographic information is available on the Internet at http://dnb.d-nb.de.

Cover image: www.purestockx.com

Publisher:
VDM Verlag Dr. Müller Aktiengesellschaft & Co. KG , Dudweiler Landstr. 125 a, 66123 Saarbrücken, Germany,
Phone +49 681 9100-698, Fax +49 681 9100-988,
Email: info@vdm-verlag.de

Zugl.: Palmerston North, Massey University, Diss., 2004.

Produced in USA and UK by:
Lightning Source Inc., La Vergne, Tennessee, USA
Lightning Source UK Ltd., Milton Keynes, UK
BookSurge LLC, 5341 Dorchester Road, Suite 16, North Charleston, SC 29418, USA

ISBN: 978-3-639-02588-0

ABSTRACT

The dairy sector is important to Argentina because it creates genuine wealth and employment. The competitiveness of Argentine dairy farms is crucial to the endurance of the dairy sector. One way to increase the competitiveness of dairy farm systems is to incorporate beneficial innovations. New Zealand (hereafter NZ) dairy systems are internationally known for their competitiveness without the presence of subsidies. Argentine dairy farmers have been attracted to NZ systems for more than 40 years. Simultaneously, NZ researchers and extension agents have been interested in extending NZ knowledge to Argentina. Despite the fact that the NZ knowledge appears to be beneficial to Argentine farms, and after so many resources spent, NZ practices have been rarely adopted. This seemingly fruitless effort in extending this technology shapes the research question of the present study: *Can Argentine dairy farmers benefit from adopting New Zealand dairy farm principles and practices?*

The main objectives of the research are the following: 1) Define a group of New Zealand ideas, practices and technologies that could be considered potentially useful innovations for Argentine dairy farmers. 2) Assess the adoption and rejection of the NZ innovations by a group of Argentine farmers. 3) Identify the reasons of adoption and rejection for each innovation. 4) Describe the impact of the adoption in the physical and financial performance of the farms. 5) Assess which have been the main causes of the non-spread of NZ innovations in Argentine dairy farms.

Seven Argentine dairy farmers, who were aware of NZ dairy systems, were selected as case studies. The data was collected through interviews, farm physical and economic records, and a field visit to the farm. In order to investigate the Argentine socio-economic environment and the Argentine dairy sector, relevant literature was reviewed and two key industry informants were interviewed. Two frameworks were utilized to analyse the qualitative and quantitative data: the Diffusion Theory (Rogers, 2003) and the IFCN network (International Farm Comparison Network www.ifcnnetwork.org), respectively.

Ten NZ innovations were defined; they were principles and practices considered typical in NZ dairy farms and not common in Argentine dairy farms. The innovations were related to four areas of the dairy system: pasture management, herd management & genetics, farm structure & organization, and human resources. The seven farmers differed in the level of adoption or rejection of the innovations.
 The two innovations most adopted were: Focus on Production per Hectare and NZ Style of Milking Shed and Milking System; and the two least adopted were: Less than 15 cows per Set of Teat-cups and other innovations related to labour productivity and Utilization of Formal Pasture Budgets. Some associations were found between the level of adoption of NZ innovations by the case study farms, the most relevant follows: increments in Return on Investment (ROI); reduction of land costs per kg of milk produced and increments in labour productivity. The NZ principle Less than 15 cows per Set of Teat-cups was found to be the innovation most closely associated with increase in labour productivity. NZ Genetics cows were found to be necessary for the adoption of seasonal calving. An association was found between the adoption of NZ Genetics and higher milk yield per kilogram of live weight, and lower mortality and replacement rates, than those that had not adopted.

ii

ACKNOWLEDGEMENTS

Thanks to the NZAID program for granting me the scholarship that made it possible for me to come to New Zealand to study. Without it, the achievement of this Master's would have been much more difficult.

I cannot find the words to thank sufficiently my supervisors Nicola Shadbolt and Colin Holmes. Nicola, I am deeply grateful for sharing your knowledge and time with me. I want to express gratitude to you for always being so positive and for believing in me. Professor Holmes, what an honour for me to have the opportunity to work with you, I will never forget your example of commitment to your work, to your students, and to the NZ dairy industry. Thank you both for clarifying and refining my thinking, rough ideas were converted with your effort into academic writing. In Argentina, I would want to show my gratitude to Bernardo Ostrowski, the IFCN coordinator for Argentina.

I would like to acknowledge the seven Argentine dairy farmers that participated in the present research. Thank you for your trust and for providing me with data from your farms, thank you also for receiving me so well in your farms and for sharing with me part of your experience and wisdom.

I would also like to acknowledge the staff at Massey University for this wonderful two years of living in Palmerston North. Especially thanks to Susan Flynn from the International Student's Office, and to Yvonne Parks and Matthew Levin from the Institute of Food Nutrition and Human Health.

To my friends from the *Latin America Society for the Development of NZ Dairy Industry* Hector Laca-Viña from Uruguay and Rene Pinochet from Chile. Thank you for those passionate discussions about dairy farmers, and their farms' productivities, risks and returns.

Thanks to you Gonzalo Tuñon and to you Javier Baudracco, for reading my drafts and for providing valuable alternative views of NZ and Argentine dairy systems and dairy farmers. Thank you for sharing the same passion for dairying, but thank to you the most for being such good friends.

Alan & Carola, Ana, Vicky and Maca, you are incredible people that have made the word "friendship" more meaningful. Thank to you for enriching our perception of the world with your different views. We feel lucky that God crossed our paths. My parents and brothers and sisters back at home, I missed you all a lot. Dad I am especially grateful to you for your constant support, for introducing me into the fascinating world of dairy farming, and for initiating me in the adoption of NZ principles in Argentine farms.

Finally, my greatest gratitude and appreciation goes to my brilliant and beautiful Maria Elisa. Thank to you Eli for helping me so much by correcting patiently my grammar, and for your constant support and encouragement during these two years of intensive experience. I am so happy that I married to you; life would be meaningless without you. Lastly, thank to you God for the gift of our baby that is still to be born.

Table of Contents (in brief)

Table of Contents (in detail)

List of figures

List of tables

1   INTRODUCTION

1.1 Background

The Spanish conquerors brought the first bovines to the humid pampas in the 1560's
and rapidly they multiplied in those fertile plains (Luna, 1994/95). The Argentine dairy
sector grew and developed, and since the very beginning has been producing,
elaborating and selling milk products for the local market without the need of subsidies
(SAGPyA, 2004a).

The following statistics show the importance of the dairy sector to the Argentine
economy: Agriculture in general provides 10% of the total employment and produces
6% of the total GDP of Argentina[1] (IFCN, 2002). Milk, is the fifth agricultural product
representing 8% of the total value of agricultural primary products after soybean, beef,
wheat and maize (SAGPyA, 2000). In the industrial phase the dairy companies are in
third place among the food and beverages industries, accounting for 11% of the total
value produced by this sector (INDEC, 2003). The Argentine dairy sector has a
domestic focus and exports only when there is an over-supply in the local market
(Gutman, Guiguet, & Rebolini, 2003), during the 1990's an average of 10% of the total
volume of milk produced was exported. Even though the sector is not focused on
exporting is a competitive sector that creates genuine wealth and employment.

Historically, milk price paid to farmers in Argentina has been decreasing (calculations
based on Gutman et al., 2003); this decreasing tendency is consistent with what was
called the *farm problem*[2]. In order to remain competitive[3], Argentine dairy farmers have
been constantly developing their production systems by adopting new technologies.
New technologies (and innovations in general) for the Argentine dairy production sector
are usually the result of research done both within and outside the country.

---

[1] In contrast, for example, in *Germany* Agriculture employs 3% of labour and generates 1% of the GDP;
in the *United Kingdom* Agriculture employs 1% of labour and generates 2% of the GDP and in the *United
States* Agriculture employs 3% of labour and generates 2% of the GDP (IFCN, 2002).
[2] The *farm problem*[2] is a theory that, in brief, states that in the long term the value of farm products tend
to decrease relatively to non-farm products (Ritson, 1977, p.145).
[3] *Competitiveness* is defined by Harrison & Kennedy (1997, p.16) "as the ability to profitably create and
deliver value at prices equal to or lower than those offered by other sellers in a specific market".

Argentina has both public and private institutions whose main objective is to generate useful innovations for its farmers, however some innovations are imported from other countries. The United States (US) is probably the country of origin of the biggest proportion of foreign innovations adopted in the Argentine dairy systems. One example of this is the high influence of US genetics in the Argentine Holstein (Molinuevo, 2001) that is the most common cow in Argentine farms (IFCN, 2002). Another example is the fact that some US companies are well settled in Argentina and are investing in research and development, and are promoting their products. Additionally, some of the most renowned Argentine specialists and researchers in dairy, studied in the US. However not all the foreign innovations come from the US. Argentine farmers have also adopted innovations from other countries including Germany, Canada, France, Australia, New Zealand and many others.

This study focused on innovations from New Zealand, mainly for three reasons: Firstly because even though New Zealand and Argentina are very different countries (in economic development, cultural background of their people, size and topography); they have some important things in common (both countries are in the southern hemisphere, have low population densities, are able to feed animals with good quality grass all year round, and for both of them the export of unsubsidised agricultural products constitutes a significant portion of their economy). The second reason is that New Zealand dairy sector is the leading exporter of milk and milk products in the world (USDA, 2004a) and can be taken as an example of coordination and efficiency for the Argentine dairy sector. The third reason, is that New Zealand farmers are considered to be among the most competitive in producing milk without the help of subsidies and they traditionally had achieved higher physical and economic performances than Argentine farmers (IFCN, 2002).

Table 1: General, Agricultural and Dairy Sector Data for Argentina
and New Zealand

	Argentina	New Zealand
General		
Population (Mill.)	36.2	3.7
Area (km^2)	2,791,810	268,021
Population density (inhab./km^2)	13	14
Total GDP (bill US-$)	264	51
GDP/capita (US-$)	7,041	13,754
Life expectancy (years)	75	79
Infant mortality (per 1000 births)	20	6
Adult literacy (%)	97.0	99.9
Agriculture (% of total)		
Land	62%	64%
Labour	10%	10%*
GDP	6%	8%**
Dairy Sector		
Dairy Cows (Mill.)	2.1	3.7
Milk Processed (Mill. kg MS)	640	1,107
Milk Exports (Mill. US-$)	280	1,710
Milk Exports (% of total volume)	11%	90-95%

Mainly for 2000 and 2001, but also for 1995 * and 1999 **.
INDEC, World Bank, Euromonitor (2004d), IFCN (2002), SAGPyA and APL. NZ Statistics (2002), NZ Statistics (2003a), NZ Statistics (2003b), Euromonitor (2004e), IFCN (2002), OECD and LIC (2002/03).

Argentine dairy farmers have been interested in New Zealand dairy systems for many decades. As a result of these interests many renowned New Zealand researchers were invited to Argentina (for example Dr. Campbell McMeekan in the late 1960's and in the 1990's Professor Colin Holmes, Professor John Hodson and Kevin MacDonald). In addition some Argentine farmers, dairy consultants and researchers visited New Zealand with the intention to better understand their systems and evaluate what can be adopted in Argentine dairy farms. As a consequence of all this the adoption of New Zealand innovations began.

1.2   The Problem, the Question and the Objectives

Summarizing the background, it can concluded that the dairy sector is important for Argentina because provides employment and wealth to the country. Dairy farms are at the beginning of the dairy chain and their survival is heavily related to the endurance of the dairy sector. Milk prices to farmers tend to decrease, so dairy farmers have to constantly develop their dairy production systems to stay in business. One way to increase their competitiveness is to adopt advantageous innovations, and New Zealand dairy systems because they are known for their competitiveness at low prices can provide useful innovations for Argentine dairy farmers.

However despite the fact that the New Zealand knowledge appears to be beneficial to Argentine farms, and that for more than 40 years Argentine dairy farmers has been interested in New Zealand systems and therefore many efforts have been done in order to extend the New Zealand principles in Argentina; New Zealand practices have been rarely adopted in Argentine dairy farms. This fruitless effort in extending this technology, make some farmers and researchers wonder:

Can Argentine dairy farmers benefit from adopting New Zealand principles and practices?

The main objectives of the present research were:

1) Define a group of New Zealand ideas, practices and technologies that could be considered as potentially useful innovations for Argentine dairy farmers.

2) Assess the adoption and rejection of New Zealand innovations by a group of Argentine dairy farmers.

3) Identify the reasons of adoption and rejection of each innovation.

4) Analyse the impact between the adoption of New Zealand innovations and the performance of the case study farms.

5) Assess which have been the main causes of the non-spread of New Zealand innovations in Argentine dairy farms.

2   THEORETICAL FRAMEWORK

2.1   Diffusion of Innovations

The theory of Diffusion of Innovations is the basic framework of the present study. Not only was the theory utilized for the data analysis but it was also crucial in elaboration of the research question and main objectives. This theory has more than 100 years of history that are presented in the first part of this section. The main elements of the diffusion theory are presented in the second section: the process of communication of a new idea through a community, and the factors affecting the speed of that communication. The last part of this section considers the main criticisms and limitations of this theory.

"An *innovation* is an idea, practice, or object that is perceived as new by an individual or other unit of adoption" (Rogers, 2003, p.12). One example of an innovation is a change to seasonal-calving on a farm that has always used an all-year-around-calving system; another example might be to start monitoring the pasture cover of a farm on a regular basis. Further, an innovative object can be, for example, to change the milking plant from a herringbone to a new rotary.

"*Diffusion* is the process in which an innovation is communicated through certain channels over time among the members of a social system" (Rogers, 2003, p.5). The diffusion process comprises all the stages, from the reception of a new idea by a specific community, followed by its communication across the community and including the process that is triggered when an individual first comes to know of the new idea. The diffusion process includes the classic process of "technology transfer" (or extension) and the process of "adoption of technology" (Black, 2000; Warner, 1974).

2.1.1 *History of the Diffusion Theory*

The origins of diffusion theory trace to Europe about a century ago, when sociology and anthropology were rising as new social sciences (Rogers, 2003).

The first scholar who studied the way in which new ideas spread among people was Gabriel Tarde. Tarde, a French lawyer, judge, and academic sociologist, was interested in learning why some innovations spread among society while others did not. The results of this first study were published in 1903 in a book called *The Laws of Imitation* (Valente & Rogers, 1995).

In his book Tarde didn't used the word adoption, he used "imitation" instead. Tarde intuitively identified important research issues that were later to be studied in a more quantitative way by successive diffusion scholars. Among other things, he observed that the rate of imitation of an innovation usually followed an S-shaped curve over time. He recognized that when the opinion leaders of a community used the innovation, acceleration in the rate of adoption occurred. Another "law of imitation" that he proposed was that the more similar an innovation is to the ideas that have already been accepted, the more likely it is that the innovation will be adopted (Rogers, 2003).

Soon after the time of Gabriel Tarde a group of British and German anthropologists undertook diffusion research. Although these anthropologists were not influenced by Tarde's writings, they studied the same phenomenon. These scientists were the first to use the word "diffusion". They are known in sociology as the "European diffusionists" because they claimed that social change could be explained by diffusion alone. This was an extreme claim because today we know that social change is caused by both invention and diffusion, which usually occur sequentially (Rogers, 2003).

The scholars who followed the work of the European diffusionists most directly were anthropologists, especially from the United States, who began to investigate the diffusion of innovations in the early 1920's. Indirectly, these anthropological schools influenced the investigation that provided the basic framework for the diffusion model, the "Iowa Hybrid Seed Corn Study" of Ryan and Gross (Rogers, 2003).

The "Iowa Hybrid Seed Corn Study" of Ryan and Gross, in 1941, is a landmark moment in the creation of the theory. Bryce Ryan, a sociologist who designed the study, was attracted to investigate non-economic factors in farmers' economic decisions. Neal C. Gross, appointed by Ryan as research assistant for the study, personally interviewed several hundred farmers in two Iowa communities. Gross asked each Iowa farmer when he decided to adopt the innovation, what were his communication sources/channels, and how much of the respondent's corn acreage had been planted in hybrid seed each year after the first trial. In addition to these recall questions about the innovation, the Iowa farmers were asked about their formal education, age, farm size, income, travel to Des Moines (closest city), readership of farm magazines, and other variables. These were later correlated with innovativeness, measured as the year in which each farmer had adopted hybrid seed (Rogers, 2004).

The result of the study was analysed and reported in three publications. The most widely known is the 1943 journal article in *Rural Sociology*, which is known as the founding document for the research specialty of the Diffusion of Innovations (Rogers, 2004). The main findings were the following:

> Ryan and Gross (1943) found that the rate of adoption of hybrid seed corn formed an S-shaped curve over time. Earlier adopters were characterized by larger-sized farms, higher incomes, and more education, and they made more trips to Des Moines, Iowa's largest city, about 75 miles away. A key finding from the seed corn study was the importance of neighboring farmers in convincing an individual to adopt the innovation (Rogers, 2004, p.15).

Figure 1 shows the adoption (cumulative and per year) of the hybrid seed by Iowa farmers from 1927 to 1945.

Figure 1: Adoption of Hybrid Seed Corn in Two Iowa Communities

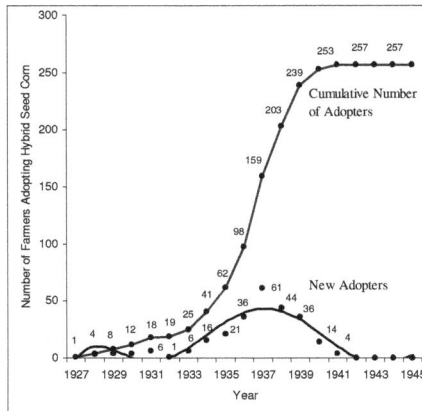

(Ryan and Gross, 1943 adapted from Rogers, 2003, p.273)

Then 10 years went by without any publication in diffusion research, in part due to the interruption of World War II. But a few years later an "invisible college" in diffusion research started to be formed (Valente & Rogers, 1995) with rural sociologists

Wilkening, Lionberger, Beal and Bohlen (Rogers, 2004). Everett Rogers joined this invisible college of scholars in 1952. After reviewing the literature about diffusion of different innovations in several environments, he argued that:

> Diffusion was a general process, not bound by the type of innovation studied, by who the adopters were, or by place or culture. I was convinced that the diffusion of innovations was a kind of universal micro-process of social change (Rogers, 2004, p.16).

In 1954, Beal and Bohlen gave their first presentation on the diffusion of agricultural innovations to the staff of the *Iowa Extension Service*. The presentation was focused on the sources/channels of communication used at various stages in the individual-level innovation-decision process, and on characteristics of farmers who adopted relatively earlier and later in the diffusion process. Soon the researchers were asked to give their presentation to audiences outside the agricultural sector. Clearly discussion of a more general model of diffusion was starting (Rogers, 2004).

In 1960, Lionberger published the first book about adoption of new ideas and practices within the rural environment. An individual-level innovation-decision process was suggested and the factors that affect the adoption and the diffusion of new ideas and technology were analysed. In 1962 Rogers published a book, *Diffusion of Innovations* where he proposed a general diffusion model for the first time, and a more standardised way of adopter categorization was developed. The book emphasized the term "innovation" instead of the numerous terms that had been used for this concept (Rogers, 2004). Ever since Rogers the several editions of this book have been the synthesis of the "diffusion paradigm". In the last 42 years more than 5000 diffusion papers, in numerous science disciplines, have been published.

In Europe van den Ban from the Netherlands, inspired by Rogers, was the first one to carry out diffusion research, in 1963 (Röling, 1988).

In conclusion it can be said that the diffusion model is a classic theory that describes a human phenomenon. This intrinsic human process, even though first described in the rural environment, proved to be a general process that displays consistent patterns and regularities, across a range of conditions, innovations, and cultures (Warner, 1974).

In the following three sections some aspects of the diffusion theory relevant to the present study are outlined: (1) the innovation-decision process model, (2) the factors that influence the rate of adoption, and (3) the innovativeness and the adopter categories. Then, some criticisms of the diffusion theory and the problem of non-adoption are addressed.

2.1.2 Innovation-decision process

An individual decision to adopt an innovation is not an instant act. Rather, it is a process that occurs over time and consists of a series of different actions. The adoption process is part of a broader procedure that is defined as the *innovation-decision process*. During this process, an individual decides to adopt or reject a new idea. It can also be said that the innovation-decision process is a particular decision-making process in which the decision is the adoption or rejection of a new idea. Based on the studies of renowned scientists such as the German psychologist Wundt, Dewey and Mead, the rural sociologists Beal, Rogers and Bohlen first posited the idea of the innovation-decision process (Rogers, 2003). The model synthesized in Rogers' book consists also of five stages[4]:

> *Knowledge* occurs when an individual (or other decision-making unit) is exposed to an innovation's existence and gains an understanding of how it functions.
>
> *Persuasion* occurs when an individual (or other decision-making unit) forms a favorable or unfavorable attitude towards the innovation.
>
> *Decision* takes place when an individual (or other decision-making unit) engages in activities that lead to a choice to adopt or reject the innovation.
>
> *Implementation* occurs when an individual (or other decision-making unit) puts a new idea into use.
>
> *Confirmation* takes place when an individual seeks reinforcement of an innovation-decision already made, but he or she may reverse this previous decision if exposed to conflicting messages about the innovation. (Rogers, 2003, p.169)

---

[4] A similar model of five stages appears in Lionberger's (1987, cited in Guerin & Guerin, 1994) called the individual-decision-process. Lionberger was part of the same invisible college of diffusion scholars and consequently agrees with the majority of the concepts. There have been other authors that developed similar models such as Chamala (1987, cited in Guerin & Guerin, 1994) and Sinden & King (1990) that vary in their details but recognise a decision process that goes through different stages, which is the most important factor of extrapolation of the classical model.

Figure 2: A Model of the Innovation-Decision Process

Communication Channels

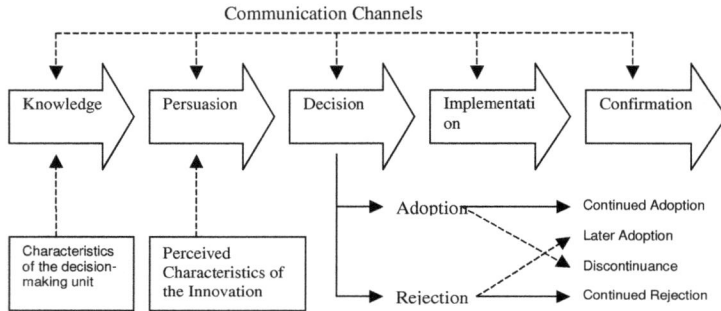

(Adapted from Rogers, 2003, p. 170)

The behaviour that occurs at each of the stages in the innovation-decision process described by Rogers (2003) is as follows:

2.1.2.1   The Knowledge Stage

The knowledge stage begins when an individual is exposed to an innovation's existence. There are three types of knowledge about an innovation:

- *awareness-knowledge,*

- *how-to-knowledge,*

- *and principles-knowledge.*

Awareness-knowledge is the information that the innovation exists. This type of knowledge may motivate an individual to get to know the innovation better. Usually, the information seeking of the second and third knowledge types occurs at the knowledge stage but it may also occur at the persuasion and decision stages. *How-to-knowledge* is the information necessary to use an innovation properly and *principles-knowledge* is the information dealing with the functioning principles underlying how an innovation works. For example, the biology of plant growth is the know-how-knowledge, which underlies the adoption of fertilizers by farmers.

To know about an innovation is different from using it. Most people are aware of many innovations that they have not adopted. A person may be aware of a new idea but not consider it relevant, or useful for his or her situation. A person's attitudes or beliefs about an innovation affect his or her way through the stages of the innovation-decision process. If the individual does not define the information as relevant to his or her situation, or if not sufficient knowledge is obtained, the process does not go beyond the knowledge stage.

2.1.2.2   The Persuasion Stage

After becoming aware of an innovation, a person may form a favourable or unfavourable attitude towards the innovation. "*Attitude* is a relatively enduring organization of an individual's beliefs about an object that predisposes his or her actions" (Rogers, 2003, p.174). While the mental activity at the knowledge stage was mainly of knowing, the main type of thinking at the persuasion stage is essentially affective. In order to take an attitude toward an innovation the person actively seeks information about the new idea, chooses what messages are considered credible, and decide how to interpret the information received. During this stage the perceived attributes of the innovation such as its relative advantage, compatibility, and complexity are especially important. In many cases, the individual mentally applies the innovation to his or her specific situation as a way of developing attitudes towards the new idea.

The main outcome of the persuasion stage is a favourable or unfavourable attitude towards the innovation. It is assumed that such persuasion will lead to adoption or rejection of the idea. But in many cases attitudes and actions may be unrelated.

2.1.2.3   The Decision Stage

The decision stage is the period when a person engages in activities that lead to the choice to adopt or reject an innovation. "*Adoption* is a decision to make full use of an innovation as the best course of action available. *Rejection* is a decision not to adopt an innovation" (Rogers, 2003, p.177). During this stage it is common for the person, to try out the new idea on a partial basis, to cope with the inherent uncertainty of the innovation. Most innovations can be tried to a greater or lesser extent, but some of them that cannot be tried partially but must be adopted completely. Usually, when a person tries an innovation and that innovation proves to be beneficial in his or her situation, the innovation is adopted.

2.1.2.4   The Implementation Stage

Up to this stage the innovation-decision process has been mainly a mental exercise of thinking and deciding. However implementation involves the person applying the innovation in practice. Typically during this stage more uncertainty arises, for example questions such as: *Where can I obtain or learn about the innovation? How do I use it? What operational problems am I likely to encounter, and how can I solve them?* Consequently, during this stage the adopter actively seeks information. Here help from previous adopters, extension agents, or sales agents play an important role.

During the implementation stage re-invention or adaptation occurs. *Re-invention* can be defined as the degree to which an innovation is changed or modified by the user in the process of its adoption and implementation (adapted from Rogers, 2003, p.181). Some innovations can be more easily adapted than others. Frequently, the innovations that are ideas and processes (software) are more adaptive than the innovations that are artefacts (hardware). In addition, a higher degree of re-invention leads to a faster degree of adoption of an innovation.

2.1.2.5   The Confirmation Stage

The confirmation stage is the period when the person seeks reinforcement of an innovation-decision already made, but he or she may reverse this previous decision if exposed to conflicting messages about the innovation. At this stage the person usually tries to avoid a state of dissonance, or if unavoidable, to reduce it. Dissonance, broadly speaking, is an uncomfortable state of mind in which the person thinks in one way and acts in a different way. A dissonant person is motivated to reduce this condition by changing his or her knowledge, attitudes or actions. Consequently if a person finds new information that makes him or her aware that the innovation is no longer beneficial to his or her situation, he or she will move towards discontinuing the innovation. Another option is to refuse to consider the new information and to simply confirm the innovation (Albrecht et al., 1989; Mook, 1987 cited in (Guerin & Guerin, 1994).

The innovation-decision process can lead as logically to the decision to reject as to the decision to adopt. In any stage in the innovation-decision process (knowledge, persuasion, decision, implementation and confirmation) the person can decide to reject the innovation. Even in the implementation and confirmation stages the person can decide to reject or discontinue the innovation.

2.1.3 Factors Influencing the Rate of Adoption

After briefly describing the innovation-decision process, the factors that influence the rate of adoption are now reviewed. *Rate of adoption* is "defined as the relative speed with which an innovation is adopted by members of a social system" (Rogers, 2003, p.219). It is usually measured as the number of persons who adopt a new idea in a specified period, such as a year (see Figure 1). Some innovations are adopted more rapidly than others. For example, the Internet was adopted very quickly by millions of people in the entire world after it publicly appeared in 1989. On the other hand some very basic agricultural innovations have been present for many years in developing countries and have been hardly adopted (Chambers & Jiggins, 1987; Guerin & Guerin, 1994).

In this section five factors that influence the rate of adoption are presented. These factors were suggested by the diffusion paradigm and with small modifications are widely accepted (see Arnon, 1989; Guerin & Guerin, 1994; Rogers, 2003; Röling, 1988). These factors are:

- Perceived attributes of innovations

- Type of innovation-decision

- Communication channels (mass media or interpersonal)

- Characteristics of the social system

- Extent of change agents' promotion efforts

2.1.3.1 Perceived Attributes of Innovations

Several studies done in Sweden[5], India[6], Canada[7] and in several states of the US[8] during the 1960s have found that the perceived attributes of an innovation explain most of the variance (between 49 to 89 percent) in the rate of adoption of an innovation (Rogers & Shoemaker, 1971). Based on this information it is expected that the perceived attributes of the innovations are the main variables that determine the rate of adoption. Consequently, is possible to predict the rate of adoption of innovations based on how the potential adopters perceive the innovations. This kind of study is called "acceptability research" and is commonly done in marketing of consumer products (Zaltman, 2003).

Rogers (2003) suggested five attributes (relative advantage, compatibility, complexity, trialability, and observability) to characterize innovations in general. However for each innovation a different group of attributes can be chosen to better characterize the innovation (Goldman, 1994; Moore & Benbasat, 1991; Murthy, Dudhani, Jayaramaiah, Veerabhadraiah, & Sethu Rao, 1973). For the present study four of the five attributes previously mentioned were suitable for the NZ innovations. Complexity was not measured because it was considered that all the NZ innovations were relatively simple to understand.

Relative advantage "is the degree to which an innovation is perceived as being better than the idea it supersedes" (Rogers, 2003, p.229). The degree of relative advantage is often expressed as economic profitability, low initial cost, a decrease in discomfort, social prestige, saving time and effort, and/or immediacy of reward. "Relative advantage" can also be defined as a ratio between the expected benefits and the actual costs of adoption of an innovation. Most diffusion studies have found that relative advantage is the strongest predictor of an innovation's rate of adoption (Fliegel & Kivlin, 1966a, 1966b; Martin, McMillan, & Cook, 1988; Sinden & King, 1990). Therefore it is widely accepted that the perceived relative advantage of an innovation is positively related to its rate of adoption.

---

[5] 1845 farmers
[6] 387 peasants farmers
[7] 130 farmers
[8] 229 Pennsylvania farmers, 88 Ohio farmers, 80 small-scale Pennsylvania farmers

The kind of the innovation determines what specific type of relative advantage (economic, social, and the like) is important to adopters. Innovations that are adopted for business purposes (e.g. a new tractor, or the use of fertilizer in pastures) are usually be oriented towards increasing economic performance. By contrast, innovations offered to consumers could be oriented towards increasing social prestige (e.g. mobile telephones, new clothes, cars). Another aspect that could determine which specific sub-dimensions of relative advantage are more important is the characteristics of potential adopters. For example, for farmers focused on increasing the profitability of their farms usually the economic benefits of an innovation are the most important. But on smaller farms, where the owner must do most of the activities of the farm, innovations that save time and effort will also be considered advantageous.

Compatibility "is the degree to which an innovation is perceived as consistent with the existing values, past experiences, and needs of potential adopters" (Rogers, 2003, p.249). An innovation that is more compatible is less uncertain to the potential adopter and, as stated above, the innovation-decision process is in some way a process of reducing uncertainty about a new idea. Past research suggests that perceived compatibility is the second attribute in importance to predict adoption (for example Kaplan, 1999; Lievrouw & Pope, 1994; Mensch, Bagah, Clark, & Binka, 1999). An innovation can be compatible or incompatible with (1) socio-cultural values and beliefs, (2) previously introduced ideas, and (3) client needs for the innovation. For example, Australian farmers place a strong value on increasing farm production. Soil conservation innovations (such as contour farming) are perceived as conflicting with this production value and have generally been adopted slowly (Guerin, 1999). The compatibility of an innovation with previously introduced ideas also facilitates adoption, for example after introducing the hybrid corn seeds and their successful results, all other hybrid seeds were introduced more easily. Compatibility with needs is also important: if a new idea or practice is useful to solve an existing problem for the potential adopter, then innovation is more rapidly adopted (Guerin & Guerin, 1994; Tully, 1981).

Complexity "is the degree to which an innovation is perceived as relatively difficult to understand and use" (Rogers, 2003, p.257). Any new idea or practice can be classified along the complexity-simplicity continuum. Some innovations are clear in their meaning while others are more difficult for the potential adopters to understand.

Depending on the nature of the innovation, complexity may not be as important as relative advantage or compatibility, but for some new ideas, complexity is a very important obstacle to adoption (Guerin & Guerin, 1994; Lynch, Gregor, & Midmore, 2000).

Trialability "is the degree to which an innovation may be experimented with on a limited basis" (Rogers, 2003, p.258). The possibility of trialling an innovation under the potential adopter's conditions can reduce uncertainty and create confidence in the innovation. Examples of this can be found in Campbell & Junor (1992) and Dixon (1982, cited in Guerin & Guerin, 1994). Consequently it is expected that the perceived trialability of an innovation will be positively related to its rate of adoption. However not all innovations can be tried or partially adopted. It is not difficult to apply more fertilizer on a specific crop on a trial basis, because it can be applied to only part of the area and then compared with the other part. It is not possible though for a farmer with a farm calving all year round, to trial seasonal calving (all cows calve at the same time) because to really notice the advantages he or she has to fully adopt the innovation.

Observability "is the degree to which the results of an innovation are visible to others" (Rogers, 2003). Some innovations are easily observed and communicated to other people, whereas others are difficult to observe or to describe. Warner (1981, cited in Guerin & Guerin, 1994) proposed that lack of observability of results would hinder the adoption of technology. Generally innovations that are ideas are less observable than practices or objects.

2.1.3.2 Type of Innovation-Decision

The type of innovation-decision is also a variable that affects the rate of adoption of an innovation. The innovation-decision can be classified as follows (Rogers, 2003):

"Optional innovation-decisions [are] choices to adopt or reject an innovation that are made by an individual independent of the decisions by other members of a system" (p.403). The innovation-decisions that are going to be analysed in this study are all of this kind. An example of this would be Argentine dairy farmers deciding whether or not they adopt New Zealand cow genetics.

"Collective innovation-decisions [are] choices to adopt or reject an innovation that are made by consensus among the members of a system" (p.403). For example, laws in

European countries that compel farmers to use soil conservation practices to prevent pollution and erosion.

"Authority decisions [are] choices to adopt or reject an innovation that are made by a relatively few individuals in a system who possess power, high social status, or technical expertise" (p.403). For example, if a dairy company decides to collect only milk at a certain temperature, therefore farmers who want to sell their milk to the company have to buy special machinery to cool the milk.

2.1.3.3   Communication Channels

Communication channels are also an important factor affecting the rate of adoption. "A *communication channel* is the means by which messages get from one individual to another" (Rogers, 2003, p.18). The communication channels can be categorized as (1) interpersonal, in contrast to mass media and (2) local, in contrast to cosmopolitan. Cosmopolitan communication channels are those that link a person with sources outside the social system under study; for example, television that connects a farmer in a specific location of Argentina with the rest of the world. Very often mass media channels are cosmopolitan. In contrast, interpersonal channels may be either local or cosmopolitan.

The following two statements are the result of studies about communication channels and their impact on the rate of adoption (Rogers, 2003): a) Mass media and cosmopolitan channels are relatively more important at the knowledge stage, and interpersonal and local channels are relatively more important at the persuasion stage in the innovation-decision process. b) Mass media and cosmopolitan channels are expected to be more important for the early adopters, and the interpersonal and local channels more important for the late adopters.

2.1.3.4   The Characteristics of the Social System

The structure, norms and opinion leadership of the social system are characteristics that affect the rate of adoption of an innovation. "A *social system* is defined as a set of interrelated units that are engaged in joint problem solving to accomplish a common goal" (Rogers, 2003, p.23). The members or units of a social system may be persons, groups, organizations or subsystems. Within the boundaries of the present study the social system is the Argentine dairy sector and the interrelated units are the Argentine dairy farms and farmers.

The diffusion theory proposes that in a specific social system there is a regular pattern in the way information flows. This pattern is called the *structure* of the system. The structure exists because not all people communicate with every member of the system with the same frequency and manner. In contrast, individuals communicate more frequently with people who have similar values. Within this structure there are some individuals who communicate with more members of the social system than others, and this quality gives them a higher degree of opinion leadership. Further there are also individuals who communicate with people outside their own social system and act as connectors between their system and others systems (Rogers & Kincaid, 1981). If two social systems are compared, and one has more interconnectedness than the other, it is expected (other things being equal) that the diffusion process of a specific innovation will be faster in the first one.

Apart from the structure of a social system, the norms of the system can also affect the rate of adoption of an innovation. "*Norms* are the established behavior patterns for the members of a social system. Norms define a range of tolerable behavior and serve as a guide or standard for the behavior of members of a social system" (Rogers, 2003, p.26). In some ways the norms of a system tell individuals what behaviour they are expected to perform. A social system can be oriented to change and facilitate the adoption of innovations, or it can have norms that are opposed to change and consequently impede the adoption of innovations (Kincaid, 2004; Tully, 1981).

The most innovative members of a system are very often perceived as people that "think out of the box" and do not completely consent to the way of thinking, values, beliefs and norms of most people within the social system. Therefore, as they are considered to be different, innovators usually do not have high influence on other people's decisions regarding adopting new practices and technologies (Rogers, 2003). However, as mentioned before, there are other individuals who communicate with more people than the average person of the system. What these people think about a new innovation is communicated more rapidly to other members of the social system. Consequently, they function as opinion leaders within the system. *Opinion leaders* are people that influence other individual's attitudes more frequently (Röling, 1988; Veerabhadraiah, Sethu Rao, & Dwarakinath, 1973a). Opinion leadership is earned and maintained by the individual's technical competence, social accessibility, and conformity to the system's norms. Opinion leaders are habitually in conformity with the system norms; therefore, when the social system is oriented to change, the opinion leaders are more innovative, but when the system's norms are opposed to change, the behaviour of the leaders also reflects this norm (Tully, 1981).

The idea of extending new technologies through the relationship between opinion leaders and extension officers was defined by the diffusion research in the early 1960's and it is still in use by extension agencies around the world (Röling, 1988). This strategy is also known as "the progressive farmer strategy".

2.1.3.5   Extent of Change Agents' Promotion Efforts

When change agents (also called extension workers or consultant officers) participate in the diffusion of an innovation, they have an effect on the rate of adoption. It has been found that the change agent's effort in diffusing the innovation is positively related to the rate of adoption (Veerabhadraiah, Sethu Rao, & Dwarakinath, 1973b). Also the empathy and credibility of the change agent with the clients can have a positive impact in the adoption of the innovation that the change agent is proposing (Rogers, 2003).

2.1.4 Innovativeness and adopter categories

Not all the members of a social system adopt a new idea at the same time; adoption occurs sequentially over time. Consequentially, based on when individuals start using a new idea, they can be classified into adopter categories. People adopting a specific innovation earlier are considered to be more innovative than individuals adopting it later. *Innovativeness*, is "the degree to which an individual (or other unit of adoption) is relatively earlier in adopting new ideas than other members of a system" (Rogers, 2003, p.96-97).

As shown in Figure 1, in the study of the hybrid corn seed in Iowa, the adoption rate of an innovation through time has a normal distribution. Based on this distribution, adopter categories can be defined (see Figure 3).

Figure 3: Adopter Categorization on the Basis of Innovativeness

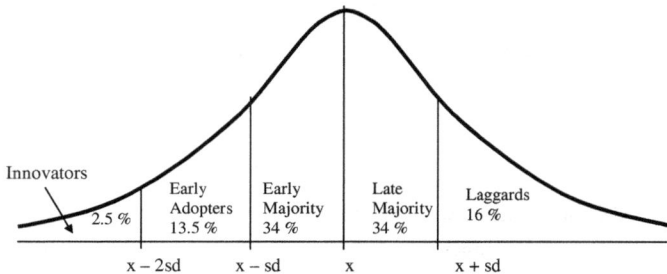

Innovators	Early Adopters	Early Majority	Late Majority	Laggards
2.5 %	13.5 %	34 %	34 %	16 %

x − 2sd x − sd x x + sd

(Adapted from Rogers & Shoemaker, 1971, p.182)

Typical characteristics have been defined for each category based on numerous diffusion studies in the last 40 years. Rogers (2003) combined these typical characteristics and defined ideal typologies of adopters:

Innovators are adventurous people; to innovate is almost an obsession with them. Their interest for new ideas could lead them out of the local circle into more cosmopolitan social relationships. Usually innovators have enough financial resources to cope with the cases when the innovation does not work. They are individuals that are able to understand and apply complex technical knowledge and can deal with high degrees of uncertainty. While other members of a social system may not respect innovators, they play an important role introducing new ideas.

Early adopters are more local people than innovators. In contrast with the innovators, they do not differ from the norms of the social system. They usually have the highest degree of opinion leadership in most systems. These early adopters are usually respected by other members of the social system and have a big impact on further diffusion of a new idea (2000).

The people that fall within the **early majority** category adopt new ideas just before the average member of a system. They interact frequently with their peers but infrequently hold strong positions. They usually get awareness of new ideas quite soon, but deliberate for a long time before adopting.

The **late majority,** together with the early majority category, are the most numerous, summing up one third of all adopters (see Figure 3). The late majority adopt just after the average person of the system. They may adopt as the result of economic necessity and of increasing peer-pressure. They are strongly influenced by the norms of the system and they usually adopt when most of the uncertainty about a new idea has been removed. They usually approach innovations with a sceptical and cautions air. They often have limited resources so that they cannot afford the luxury of adopting disadvantageous innovations.

The **laggards** are the last ones in a social system to adopt a new idea. They are very local and some of them are not very social (quite isolated). Their point of reference is the past and they are very traditional. They tend to be suspicious of innovations and change agents. Their resistance to innovations could be entirely rational from their point of view, as their resources are usually limited and they must be assured that a new idea will not fail before adopting it.

It is important to note that the categories described above are ideal and typical. They do not represent individual farmers; they describe individuals with extreme characteristics. Howden & Vanclay (2000), in their research about "mythologization" of farming styles in Australia concluded, that although several focus groups identified several farming styles, when they tried to validate the styles with real farmers they could not find farmers that fulfilled all the main characteristics. The same would happen with adopter categories, because no real farmers match the categories of adopters. The categories are only a simple way to describe the characteristics that make a person more or less an innovator.

2.2 Criticisms of and Responses to Diffusion Theory

Having outlined the main points of the diffusion theory, a consideration of the main criticisms to this framework is presented.

2.2.1 Pro-Innovation Bias

The pro-innovation bias is one of the most serious shortcomings of diffusion theory. This was the first bias to be recognized by diffusion scientists:

"The *pro-innovation bias* is the implication that an innovation should be diffused and adopted by all members of a social system, that it should be diffused more rapidly, and that the innovation should be neither re-invented nor rejected" (Haider & Kreps, 2004, p.7).

The diffusion theory has been developed based on innovations that were successfully diffused and adopted by most of the members of a social system. In most of those studies, the innovations were clearly advantageous and had a positive impact on the situation of the adopters. The implicit idea behind this theory is that to adopt innovations is always beneficial and that to reject new ideas and technologies is irrational.

The diffusion theory also implied that all members of a social system should adopt the innovation. It does not consider the different situations and characteristics of members of the same social system (Arnon, 1989; Röling, 1988). An innovation could be beneficial for some people and useless for others; in those cases where the innovation is not beneficial, the most intelligent decision for those people would be to reject the innovation. For a third group of members of a social system the innovation has to be re-invented or adapted to be advantageous. The concept of re-invention was added later to diffusion theory; in the beginning the only two possibilities were to adopt or to reject an innovation.

Change agents and researchers affected by the innovation bias have participated in the adoption of several practices that were detrimental to the communities that adopted them. The adoption of those innovations had a negative long-term impact on the communities that adopted them (Black, 2000). Examples of this are contamination of streams by excess usage of fertilizers and the salinization of soils caused by inappropriate irrigation.

Several ways were suggested by Rogers (2003, refer to p.112) and Haider & Kreps (2004, refer to p.8) to overcome this bias. Both studies agree that the most important step is to recognize that this bias could exist and to consider that sometimes the best option is to reject an innovation. Further, it was suggested the potential consequences of the adoption of new practices be carefully studied before extending them.

2.2.2 Person-Blame Bias

Most of the diffusion research has been done from the point of view of the developers of the innovations and the change agencies (also known as the *sources* of the innovations). This bias is probably suggested by the words that are used to describe this field of research: "diffusion research" might equally have been called "problem solving", "innovation seeking", or "evaluation of innovations" from the point of view of the potential adopter (Rogers, 2003).

The fact that the point of view of the developers has been present in most diffusion studies has put the blame on the potential adopters (also called the *audience* of the innovations) when the level of adoption of the innovation has been low. This orientation implies that: *If you do not want to adopt the innovation there is a problem with you.* Instead would be more appropriate to put the blame on the innovation itself (may be the innovation is not beneficial for that community) or in the way in which the innovation was communicated (Caplan & Nelson, 1973).

As in the pro-innovation bias, Rogers (2003, refer to p.125) and Haider & Kreps (2004, refer to p.8) suggested several ways to overcome the person-blame bias. Yet again both studies agree that the most important step is to recognize that this bias exists and that studies should be designed in order to avoid it. Some ways to prevent this bias are: not using individuals as sole units of analysis but consider also other aspects such as the attributes of the innovations, the intentions of the change agency or company, and the characteristics of the social system.

2.2.3 Pro-Innovativeness Bias

Past studies (previously cited) have correlated the characteristics of individuals with their degree of innovativeness. It has been found that better educated and wealthier farmers are among the most innovative. People that are considered more rational, with a more positive attitude towards science and better able to cope with uncertainty adopt before the average person of a social system. Finally it has been found that people that have more social participation, and who are more cosmopolitan, have a greater degree of innovativeness (Rogers, 2003).

These findings could induce researchers to think that the more innovative a person is, the higher the probability that a person is successful. However correlation does not necessarily imply causality. These findings cannot say that a more innovative person will be wealthier; instead they mean that there is a relationship between wealth and innovativeness without assessing the cause or if other variables are affecting the results (Meyer, 2004).

Another aspect that is important to clarify is that there are very few people who adopt early all kinds of innovations. It is reasonable to think that different people are more innovative regarding different kinds of innovations. For example, one farmer can be more innovative in respect to tractors and machinery and a second farmer can be more innovative with respect to pasture managing techniques or in computer utilities. Therefore it is an over-simplification to classify people as innovators or laggards for all kinds of innovations.

2.2.4 Heterogeneous Social Systems

The diffusion theory implies that once opinion leaders of a social system adopt an innovation then there is a natural diffusion of the innovation to the rest of the members of the social system. However, whereas most of the seminal diffusion studies were done in relatively small and homogenous communities, the definition of social system utilized within the diffusion theory is very broad and can include, for example, "all farmers in Latin America".

When the diffusion theory started to be applied in developing countries, it was observed that extension agencies communicated more often with farmers who were more responsive. These farmers were usually relatively wealthy and educated and became more wealthy and educated helped by extension agencies (Röling, Ascroft, & Wa Chege, 1981). This problem was called the inequality problem and it was thought that the theory could not be applied to social systems that were heterogeneous (Arnon, 1989; Chambers & Jiggins, 1987; Goss, 1979; Röling, 1988) because the wealthy-poor gap would be enlarged.

This inequality problem could be solved by the correct definition of the target social system. Röling at al. (1976) in Kenya, and Shingi and Mody (1976) in India (both cited in Rogers, 2003), designed diffusion approaches that narrowed, rather than widened, socioeconomic gaps. They accomplished this by targeting correctly the community in which they wanted to extend the innovations (Röling, 1988). For example, if the intention is to diffuse rural innovations among resource-poor farmers in a region, then the change agents should communicate with the opinion leaders of the resource-poor farmers of that area, and once the opinion leaders have adopted, the natural diffusion of the innovation will occur. However if change agents contact opinion leaders of wealthy farmers, what would probably happen is that the innovation would be spread through the wealthy farmers of the region and not necessarily to other groups of farmers.

2.2.5 Diffusion Theory and "top-down" Models of Extension

As previously mentioned, the classic diffusion paradigm has been the originating theory of two extension strategies, one that was called the "progressive farmer" or "opinion leader" model (Arnon, 1989; Röling, 1988) and a second one called the "top-down" or "linear adoption" strategy (Black, 2000). However the diffusion theory cannot be defined as the sum of these two technology transfer strategies. The diffusion paradigm was only the theory that inspired these two extension models. The diffusion process is a broader social process than an extension strategy. It is a human phenomenon that occurs with or without the presence of extension officers or change agencies (Warner, 1974). Consequently, the problems identified in both the "progressive farmer" and the "top-down" extension strategies are not deficiencies of the diffusion theory. They really are consequences of the implementation of these extension methodologies.

2.2.6 Participatory Approaches and Diffusion Theory

There are some innovations that do not diffuse naturally through the members of a target social system. These innovations are usually not advantageous in the short-term, not very compatible, are difficult to understand or explain, or have any other attribute that make it difficult for them to be extended (Black, 2000). The constraints could be also in the context of the social system; a volatile context with high risk and uncertainty, or a context where it is difficult to access credit, or where it is difficult to communicate the innovation, will make adoption more difficult (Arnon, 1989). There is also some evidence that showed that in more traditional (Tully, 1981) or low-educated and poor-resource communities (Arnon, 1989) it is more difficult to extend new technologies.

The failure of the "top-down" models in extending low acceptability innovations (Guerin & Guerin, 1994), or innovations with acceptable attributes in traditional or low-educated and poor-resource communities (Chambers & Jiggins, 1987), motivated the appearance of the "participatory approaches". These approaches in which scientists, extensionists and potential adopters work together in the origination and diffusion of innovations; have proved to be successful in many cases (Carberry et al., 2002; Murray, 2000).

But the fact that participatory approaches have been successful in extending some new ideas does not imply that classic methods are no longer useful. Both "top-down" and participatory approaches seem to be useful in extending innovations. They are the two extremes of the extension spectrum. It is quite accepted that the methodologies are complementary. It is also accepted that in different situations, or for diverse innovations, one of the extension strategies could be more appropriate than the other (Black, 2000; Chambers & Ghildyal, 1985; Guerin & Guerin, 1994).

2.3 Constraints on Adoption of Innovations in Agriculture

The problem of non-adoption in agriculture is common around the world (Guerin & Guerin, 1994). Very often there are ideas and practices that experts perceive as beneficial for farmers to adopt but farmers are not interested. Much research in this field has been carried out. Especially in developing countries where the need for basic agricultural technologies is great (Arnon, 1989; Chambers & Jiggins, 1987; Röling et al., 1981). There has also been research in Australia in relation to the adoption of more sustainable practices (Guerin, 1999; Lockie, Mead, Vanclay, & Butler, 1995; Vanclay, 2004) and other kinds of technology; for example, intelligent support systems (Lynch et al., 2000).

The "problem" of non-adoption is clearly multi-factorial. Guerin & Guerin (1994) in a comprehensive review of this field identified several factors that were found to constrain the adoption of innovations. Based in the factors proposed by diffusion theory the reasons for non-adoption can be classified in four groups.

These four groups relate to:

- The attributes of the innovations as perceived by the potential adopters within their own context/ situation.

- The characteristics of the social system with respect to intercommunication, opinion leaders and norms.

- The channels through which the innovation is communicated.

- Whether a change agency participates or not in the diffusion process, the extent of effort of the change agents and the extension strategy utilized by the extension agency.

A brief analysis of these four groups of constraints follows:

2.3.1 Perceived Attributes of Innovations

Multiple factors influence the way in which farmers perceive innovations. The following factors do not affect innovations directly but they affect, in contrast, the way in which farmers perceive them:

Factors Inherent in the Farmer: existing needs, socio-economic status, age, goals, family situation, level of education, personality in general, level of aspiration, attitude towards farming, attitude towards economic future of farming, attitude towards risk, attitude towards technology, attitude towards change, beliefs, values and fears.

Factors inherent in the farm or business: scale, location, and other characteristics.

Economic conditions of the context: economic benefits of the innovation within the specific context, access to credit or other way of financing to buy the innovation, prices at the international and local level, foreign exchange fluctuations, government policies, import duties and demand for food.

In relation to the innovation: perceived benefits (comparative advantage, economic benefits, status benefits, comfort benefits), initial and maintenance costs, perceived compatibility, perceived complexity, perceived trialability and perceived observability.

2.3.2 Characteristics of the Social System

As previously stated, the main factors that could limit the diffusion of an innovation are: low level of inter-connectedness and system norms resistant to change.

2.3.3 Communication Channels

The lack of resources to advertise or promote through mass media channels can limit the awareness of an innovation to only a small group. On the other hand, as most innovations cannot be fully communicated through mass media alone, lack of resources to provide advice to farmers during the innovation-decision process (specially in the persuasion and implementation stages) can slow the diffusion process.

2.3.4 Extension Agency and Agents

The presence of an institution interested in diffusing the innovation usually accelerates the rate of adoption. A change agency can provide funds to promote the innovations in the mass media and train people to give advice to farmers during the innovation-decision process. Complex innovations, those integrated or with low observability are more difficult to extend. This kind of innovations requires more contact between the change agent and the farmer (more participatory approaches) and/or a higher level of technical understanding by the farmers and/or more skilful change agents (Black, 2000).

3 DAIRY SECTORS: New Zealand and Argentina

In the next two sections the New Zealand and Argentine dairy sectors are described in order to get an overview of the environments in which the dairy farms of these two countries are immersed.

3.1 New Zealand Dairy Sector

3.1.1 Physical and Economic Foundations

There are two dominant factors that make the New Zealand dairy sector the leading exporter of dairy products (USDA, 2004a). Firstly and best known, is the production system based on pasture. New Zealand is a rough country and does not have soils that are naturally fertile. What it does have is a temperate, moist climate with regular rain that is nearly ideal for growing temperate pastures. This is the country's main natural resource. Grass has the potential to be grown all year round and grazed intensively over approximately 45% of New Zealand's area[9]. Productions of 10 to 16 tonnes of Dry Matter of high quality grass per year are achieved on New Zealand dairy farms (Holmes, 2003b). "The efficient utilization of the grass through integrated management of pasture and livestock is the economic cornerstone of the dairy industry. It is the basis of the industry's comparative advantage internationally" (Mitchell, 2002, p. 4).

The second fundamental factor of the New Zealand dairy sector is its export orientation. "No other dairy industry in the world, and few industries in any other sectors, match the focus the dairy industry in New Zealand has on international customers" (Mitchell, 2002, p.4). The domestic market is nearly insignificant. It consumes only about 5% to 10% of New Zealand milk and milk products (Statistics-NZ, 2000). The rest is sold to overseas markets thousands of kilometres away. This requires a remarkable capacity to build commercial relationships with companies and customers all over the world.

---

[9] Arable land + grassland + horticulture = 12,060,945 hectares. From this land 1,794,474 hectares (15%) are used for dairy farming (own calculations from 2002 National Agricultural Production Census).

In New Zealand the dairy sector is the leading exporting sector of the economy. In the last 6 years (from 1997 to 2003) the dairy sector earned between 16 to 20% of total New Zealand's exports by value (Statistics-NZ, 2003). In relative terms it is much more important to New Zealand than, for example, the automobile sector to Japan (Mitchell, 2002). The performance of the sector is, therefore, not just important for New Zealand dairy farmers or others that work in the sector, it is also of weighty significance to the nation as a whole.

3.1.2 Economic Environment

New Zealand's economy has performed well in recent years, with a growth rate of 3% per year from 1998 to 2003, and inflation kept under 2.7% per year (see Table 1).

Table 2: New Zealand Economy – Financial Indicators

	1998	1999	2000	2001	2002	2003	Avr	Dev (%)
Inflation (% change)	1.3	-0.1	2.6	2.6	2.7	1.8	1.8	60%
Interest rate (%)	11.2	8.5	10.2	9.9	9.8	9.8	9.9	9%
Exchange rate (per US$)	1.87	1.89	2.2	2.38	2.16	1.72	2.04	12%
GDP (% real growth)	-0.2	3.9	4	2.5	4.3	3.5	3.0	56%

(Euromonitor, 2004c)

The exchange rate between the New Zealand dollar (NZ-$) and the United States currency (US-$) has had an increasing trend since 1998. The appreciation of the NZ-$ affects negatively New Zealand's economy as it is heavily dependent on commodity exports.

The interest rates for long-term loans have been around 10% during recent years, which is one of the highest interest rates within developed countries but much lower than in many developing countries[10].

---

[10] Average interest rates for some developed countries between 1996 and 2002 have been 6.7% for France, 9.4% for Germany, 4.5% for The Netherlands, 6.9% for USA, 4.0% for Switzerland and 5.2% for UK. For developing countries for the same period have been for Argentina 22%, for Brazil 69%, for Chile 12% and for Uruguay 68% (Euromonitor, 2004b).

3.1.3 Milk Price paid to farmers

Milk and milk products prices in the international arena are much lower than in most of the domestic markets[11] (*IFCN Dairy Report*, 2003). This is mainly caused by the restrictions on imports by tariffs and tariff quotas and by direct and indirect subsidies. Prices in the export markets are also far more volatile than prices in most domestic markets because international markets cannot be controlled and because there is a great number of factors influencing them (Mitchell, 2002).

The following figure shows changes in the company payout, in real terms, paid to farmers in the last 30 years.

Figure 4: Evolution of the New Zealand Company Payout (1973/74-2002/03)

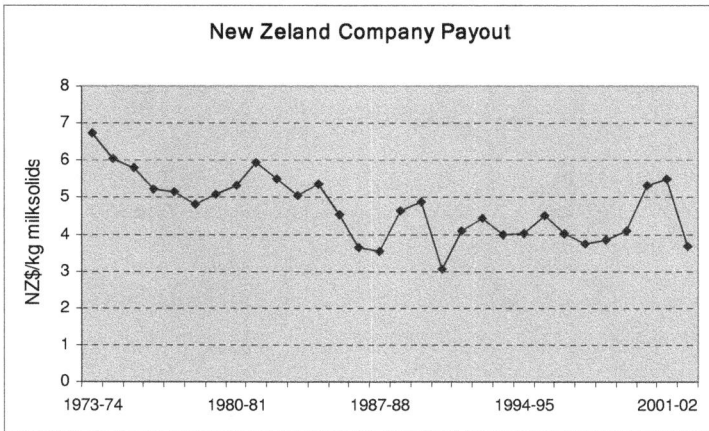

(LIC, 2002/03)

Prices weighted to give real dollar values using the Consumers Price Index for the end of the June quarter. Average Dairy Company total actual Payout for 1974/75 to 1988/89 has been derived from NZ-$/kilograms milk-fat, a relationship of 74% milk-protein/ milk-fat has been considered.

---

[11] Milk prices for Ireland, The Netherlands, Spain, Germany, France and Belgium from 1996 to 2002 ranged between 27 to 37 US-$/kilograms of standardized milk (*IFCN Dairy Report*, 2003). For New Zealand the milk price from 1996 to 2002 ranged between 13 to 20 US-$/kilograms standardized milk (*IFCN Dairy Report*, 2003).

The company payout has a decreasing trend of 1.3% annually and varied remarkably from one year to another. There have been very low prices like NZ-$ 3.04 (1990-91) and high prices like NZ-$ 5.50 (2001-02).

As previously mentioned, the payout for dairy farmers is affected by changes in commodity prices. Other important factors that influence milk price are the fluctuations in exchange rates, the product mix, and the sales volume of the different cooperatives and companies. To reduce the impact of the fluctuations on exchange rate, the main New Zealand cooperative (Fonterra) has a foreign exchange hedging policy (*Fonterra News*, 2003). This policy allows the cooperative to predict more exactly the payout for the following year and guarantee to the farmers a minimum price for their milk for the whole season, 15 months in advance.

3.1.4 Structure of the Sector

The New Zealand sector is an integrated cooperative system, made up of five main components (adapted from Holmes, 2003b):

The farmers are the owners of the whole sector (there were 13,900 dairy farmers in 2001 in New Zealand, IFCN 2003). Consequently the industry is focused on maximising their returns.

Fonterra Cooperative Group, which collects, industrializes and sells 95% of New Zealand milk all over the world. There are also some other smaller dairy cooperatives and companies, which make up for the remaining 5%.

Livestock Improvement Corporation, which operates the dairy sector's genetic improvement programme and provides essential services to dairy farmers and *Fonterra* (for example, herd testing for 2.8 million cows, artificial breeding for 2.6 million cows, sire proving service, research and development).

Dexcel (**Centre for excellence in dairying**), research company responsible for all sector-good activities, to optimise whole farm systems: extension, farm production, research, economic modelling, and strategic planning.

Fonterra Research Centre, a section from Fonterra, which provides essential services to dairy companies in relation to the manufacture of dairy products. This centre is in charge of basic research, technical and product development programmes and of all aspects of the dairy manufacturing industry.

3.1.5 Size and Growth of the Sector

During the 2002/03 seasons the 3.8 million New Zealand cows produced 1,191 million kilograms of milksolids (LIC, 2002/03). This comprises 2% of world's total milk production; however New Zealand is the main exporter country of milk and milk products in the world with more than 25% of the world's total trade in dairy products in the last 5 years (USDA, 2004b). The evolution of the New Zealand milk production in the last 30 years is shown in the following figure:

Figure 5: Evolution of NZ Milk Production (1974/75-2002/03)

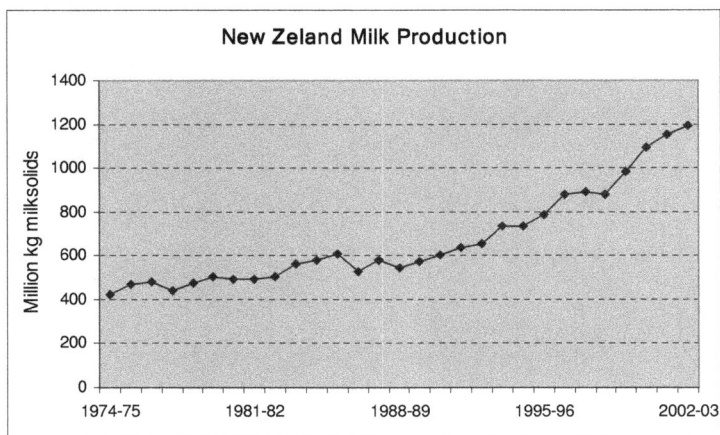

(LIC, 2002/03)

During the last 30 years in New Zealand, there has been a continuous growth of milk production. This growth has been at a rate of 3.4% annually and the total milk production has tripled from 1974 to 2003.

During the 2003/04 season a 5.3% increase in milksolids processed was registered (LIC, 2003/04), despite the very difficult climatic conditions for most regions. In the beginning of the year there were forecasts of a payout only slightly higher than 2002/03, and expectations of a further drop in 2004/05 (MAF, 2004). The final payout figure lifted throughout 2004 to NZ-$ 4.25 per kilograms of milksolids (Fencepost.com, 2004).

The number of herds decreased during this season following a trend that started several years ago. Number of cows increased by 110,000 and, as a result, the average number of cows per herd increased to over 300 cows (LIC, 2003/04).

During the next season (2004/05) milk production is expected to continue increasing, helped by better climatic conditions. The forecasts payout for farmers is NZ-$ 3.85 per kilogram of milksolids (MAF, 2004).

3.2 Argentine Dairy Sector

3.2.1 Physical and Economic Foundations

Argentine dairy production systems are based on two main sources of feed: pastures, and concentrates and silages. The temperate and relatively moist climate together with naturally fertile soils allows pastures and crops to grow all year round. Therefore, pastures can be grazed all year round and grains and silages can be bought at very competitive prices.

In contrast with New Zealand, the dairy sector in Argentina is focused on the domestic market. Only an average of 4% of the total milk and milk products in 1980's and 10% in the 1990's were exported.

The next figure shows the proportion of the total Argentine milk production that has been exported from 1990 to 2003 (14 years).

Figure 6: Argentine Exports in Percentage of Total Milk Production

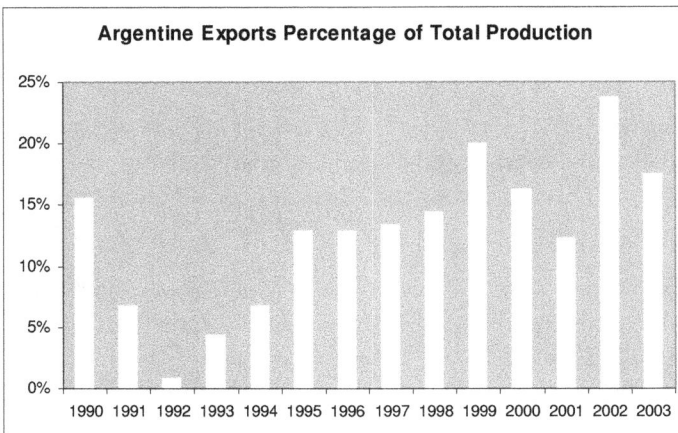

(CIL, 2004; Gutman et al., 2003; SAGPyA, 2004a)

Additionally, exports of milk and milk products are not the main exported products of the country as in New Zealand. In Argentina, dairy exports have traditionally been a very small proportion of the total exports of the country (for example, in the year 2000 only 1% of the total value of exports) (Gutman et al., 2003).

3.2.2 Economic Environment

In December 2001, after 4 years of economic recession that had a remarkable social impact, the president of the country abdicated. In January 2002 a provisional president decided to do an asymmetric devaluation of the Argentine currency, "the Peso", that had severe effects in the financial markets. An external and internal default was announced and the "Country Risk" reached one of the highest values in Argentine history. All this generated a huge uncertainty in the country and the economic recession worsened; the gross domestic product (GDP) of the country fell by 11% in 2002. A new president was elected in April 2003 and that year the economy recovered by nearly 9%.

Table 3: Argentine Economy – Financial Indicators

	1998	1999	2000	2001	2002	2003	Avr	Dev (%)
Inflation (% change)	0.9	-1.2	-0.9	-1.1	25.9	13.4	6.2	182%
Interest rate (%)	10.6	11	11.1	27.7	51.7	19.1	21.9	74%
Exchange rate (per US$)	1.00	1.00	1.00	1.00	3.06	2.90	1.66	62%
GDP (% real growth)	3.8	-3.4	-0.8	-4.4	-10.9	8.7	-1.2	584%

(Euromonitor, 2004a)

The exchange rate with the US-$ after the 200% devaluation in December 2001 has been quite constant around AR-$ 3. A strengthening of the AR-$ could be negative for raising exports of commodity products that is one of the main forces of the GDP growth.

In the coming years, the GDP is predicted to grow 5.5% in 2004 and 4% in 2005. After very high inflations rates in 2002 and 2003, it is expected to decrease to 6.7% in 2004 and increase again to 9.5% in 2005 (Euromonitor, 2004a).

Traditionally, interest rates have been high in Argentina and access to credit has been difficult in comparison with other countries. Last year (year 2003) the interest rates were close to 19%, and in 2004 there has been a decrease in the interest rates making access to credit slightly easier.

3.2.3 Milk Price paid to farmers

As was previously mentioned, Argentine milk price is mainly defined by the supply of and demand for milk in the domestic market. This market is completely deregulated and when an excess of milk is produced the price falls immediately. A peculiarity of the Argentine dairy sector is that it has a double seasonality. One is within the year in which the prices change from month to month with the highest prices during the winter and the lowest during the end of spring and the beginning of summer. There is also a second seasonality, of indefinite number of years, that begins with periods of high demand for milk and relatively high prices followed by periods of excess of supply and relatively low prices.

In the following figure changes in the milk price, in real terms, paid to farmers in the last 30 years are portrayed.

Figure 7: Evolution of the Argentine Milk Price to Farmers (1974-2003)

From 1974 to 2000 Gutman et. al., 2003; years 2001, 2002 and 2003 estimated from data collected from a group of big farms of the West of Buenos Aires Province.
All prices weighted to give real values using the Consumers Price Index (base 1999).

Milk prices had been decreasing in real terms at a rate of 4.4% annually and had remarkable variations from one year to another. The biggest changes are usually caused by deviations in the exchange rates or by major problems in the country's economy.

In the next figure the "within years seasonality" can be seen. It can be noticed how during 2000, 2001 and 2003 the milk price was high during the winter (June-July) and then decreased in the spring (October-November). During 2002 the price kept increasing due to a strong demand for milk by the milk companies.

Figure 8: Monthly Evolution of the Argentine Milk Price to Farmers (2000-2003)

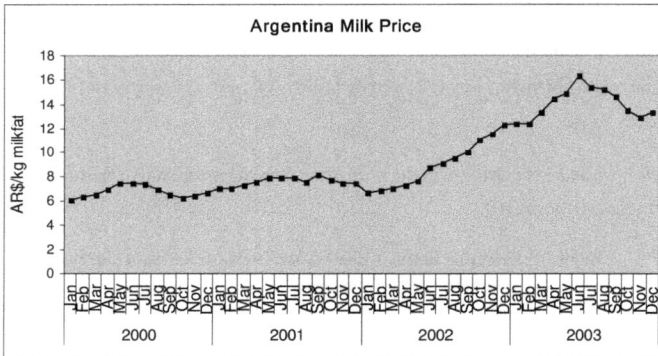

From 1974 to 2000 Gutman et. al., 2003; years 2001, 2002 and 2003 estimated from data collected from a group of big farms of the West of Buenos Aires Province.
All prices weighted to give real values using the Consumers Price Index (base 1999).

3.2.4 Structure of the Sector

The Argentine dairy sector, in marked contrast with New Zealand's, is not very integrated and does not have a common strategic plan. Even further, one of its main characteristics is a continuous conflict between the dairy companies and the farmers regarding the milk price.

The approximately 13,000 Argentine dairy farms are very heterogeneous. There are very big farms (more than 1,000 cows) and also very small farms (less than 50 cows) with an average of 174 cows per farm (Gambuzzi, Zehnder, & Chimicz, 2003). The production systems are also very different and change across the different dairy regions (Gutman et al., 2003). Another peculiarity of the Argentine dairy farmers is that they do not have an institution with enough representative power to negotiate with dairy companies and the government.

There are approximately 800 dairy companies in Argentina (Gutman et al., 2003). They can be divided into three well-defined groups. The first group includes a small number of big companies most of them Argentine owned. The 6 biggest dairy companies process 53% of the total milk production and they export most of the milk and milk products of the country. The second group includes a significant number of medium companies, each processing less than 1.5% of the total milk produced (Ostrowski & Deblitz, 2001). And the last group consists of hundreds of small companies that usually specialize in producing cheese and very often are accused of not paying taxes and not complying with hygiene regulations (Gutman et al., 2003).

Research and extension in dairy farming in Argentina is not given the same importance as in New Zealand. Argentina does not create much knowledge within its dairy sector and normally farmers and consultants look for innovations for their production systems to other countries as the United States, Australia and New Zealand (researcher's experience in the Argentine dairy sector).

3.2.5 Size and Growth of the Sector

After 8 years of an average growth of 9%, the highest milk production occurred in 1999 (10,329 million litres). However from 2000 to 2003 milk production decreased to similar volumes as in 1994. During 2003 approximately 1.2 million cows (SAGPyA, 2003) produced 7,951 million litres (approximately 538 million kilograms of milksolids). The following graph shows the Argentine total milk production from 1970 to 2003 (21 years).

Figure 9: Evolution of the Argentine Milk Production (1970-2003)

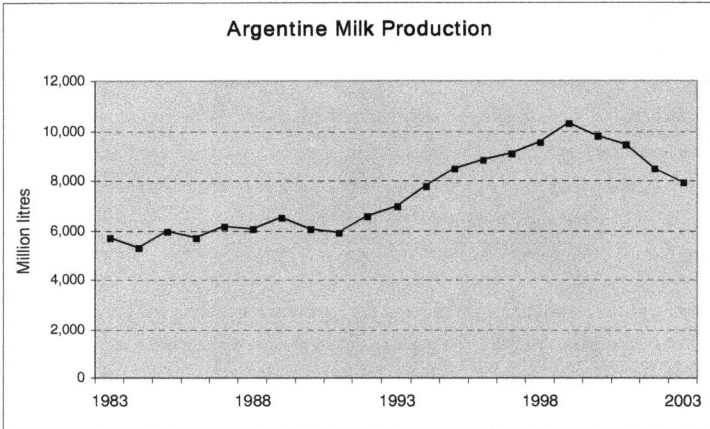

(SAGPyA, 2004a)

During the period shown in Figure 6 the annual growth in total milk production has been 3%, which is lower than the 3.4% annual growth of New Zealand.

Throughout 2004 the milk production increased approximately 20% (SAGPyA, 2004b) and a further growth is forecasted for 2005.

4 DAIRY FARM SYSTEMS: New Zealand and Argentina

Farm models were chosen from the International Farm Comparison Network (hereafter IFCN) as examples of dairy systems of New Zealand and Argentina. There are six farm models, three for each country; these models were defined as typical[12] for each country. What follows is an introduction to the IFCN framework.

4.1 The IFCN

The IFCN is a worldwide association of agricultural scientists, consultants and farmers. Most of the main countries that produce agricultural foodstuffs are represented in the network; every continent is represented. The framework is divided into sections that analyse systems of production of different agricultural products; IFCN Dairy, the IFCN Beef and the IFCN Arable Crops. In the IFCN Dairy there were 27 countries represented in the last annual report (2004) with 83 different typical farms. The institutions that originated the network are the Federal Agricultural Centre (FAL) from Germany, and the Agricultural and Food Policy Center (AFPC) from the Texas A&M University from the United States.

In brief, the questions that the IFCN wants to answer for each country are:

How is farming done (farming system, production technology)?
What is the cost of production?
What are the reasons for advantages and disadvantages in competitiveness?
What is the future perspective of agricultural production at the locations considered?

(adapted from Isermeyer, Deblitz, Hemme, & Plessmann, 2004).

To answer these questions, the IFCN utilizes the following methods and principles; here they are illustrated by reference to the IFCN Dairy for the agricultural product "milk":

The concept of "typical regional models": the first step is to select the most important regions and locations for milk production in the world. Regional experts and local farmers in those areas are contacted and a "panel" is organised to define typical farms

---

[12] Note that 'typical' means characteristic, which is not the same as most frequent (mode) or a statistical average.

for the region. This panel aims to define a moderate and a large farm for the region, but variation can be found in different regions. The idea is to establish the data for each typical farm in a very detailed manner, and as close to reality as possible. The data have to cover all relevant aspects, ranging from the physical organization of enterprises to taxation of the profit. Statistical averages derived from farm groups are avoided unless they are very accurate. In addition, real farms are not used because they have peculiarities that deviate too much from the typical (adapted from Isermeyer et al., 2004).

The "principle of total costs": in the framework not only cash expenses are taken into account but also non-cash adjustments (including the costs of depreciation, the change in inventory and gains or losses in capital value), and the opportunity costs of the factors that are owned by the farmer (capital, land and labour).

The computer model TIPI-CAL[13] is utilized to input all the physical and economic data from the different models and to compare costs and economic indicators across the different models. The model works in MS-Excel® (spreadsheet) in order to enable everyone involved in the IFCN to have access to the database. It has two main parts, one where all the actual data from each model is registered and another that simulates for periods up to 10 years, to evaluate growth, investments and the impact of changes in policy and prices.

The whole farm data is analysed. The typical models take into account the data from all the activities (or enterprises) of the farm business (for example, dairy, beef and cash crops) plus the income and the expenses of the owner (including the income from off-farm activities and family expenses). To calculate the total cost of each product, the indirect costs are distributed proportionally (for example the machinery expenses are distributed according to the amount of time used in the different activities).

The costs structure and the economic indicators are explained in further detail in the next section. For more details about the IFCN refer to www.ifcnnetwork.org.

---

[13] TIPI-CAL: Technology Impact and Policy Calculations

4.2   Typical Dairy Farms (IFCN)

The six typical dairy farm models are presented in the following way:

Firstly, two typical farms, one for each country, are described and contrasted with the average dairy farm in those countries. Secondly, the main characteristics of the two farms are compared. Finally, details of four other farms are presented in order to describe the different dairy farm systems within each country.

4.2.1 NZ-239-cows and AR-350-cows farm models

The following two farm models for New Zealand and for Argentina were selected by the IFCN to represent the farms that contribute the biggest proportion of milk for each country[14], with management levels that are average or above average[15]. However they are not average farms in any aspect (size, management level, or any other). The data is from the 2000-2001 season for the New Zealand farm and from the year 2001 for the Argentine (IFCN, 2002).

The **New Zealand model farm with 239 cows** (NZ-239) is based on data from an annual survey (*Profit Watch*) carried out annually by *Dexcel* (a research and development company owned by New Zealand dairy farmers). The average data from the South Auckland region was chosen to define the most typical New Zealand farm (Nicola Shadbolt, IFCN coordinator for New Zealand, personal communication). In the South Auckland region are 32% of all New Zealand dairy farms (LIC, 2002/03) and they produce 29% of the total New Zealand milk (calculations based on LIC, 2002/03). It is also considered that the typical South Auckland farms are similar to those of the Taranaki region, where there are a further 16% (LIC, 2002/03) of New Zealand's herds and 12% of its milk production (calculations based on LIC, 2002/03). Consequently the model defined represents a typical farm for two major regions that together contribute 41% of New Zealand milk, and contain 48% of New Zealand dairy farms. This farm

---

[14] 'Farms that contribute the biggest proportion of milk' means that each of these two farms are representative of the category of dairy farms that produce more milk in each country. In order to do this, all farms in the country were categorized by number of cows by the IFCN experts (Ostrowski & Deblitz, 2001). The category that contributes with the biggest proportion of milk in Argentina is the category of farms with 250 to 350 cows. For New Zealand, farms located in the Taranaki and Waikato regions form this category.

[15] The IFCN experts in each of the countries assessed the management levels.

model is different from the statistical average because the average for New Zealand has 251 cows, and an effective area of 96 hectares (LIC, 2000/01) (see Table 4).

Table 4: Differences in Size Between IFCN Models and Statistical Average for NZ (2000/01)

	NZ-239	Average Farm	Difference
Number of Cows	239	251	5%
Hectares	104	96	8%

The **Argentine model farm with 350 cows** (AR-350) is based on the average data of a group of farms in the Córdoba province. This farm model, and another of 250 cows located in the Santa Fe province were considered typical of the farms that together contribute the biggest proportion of the Argentine total milk production[16] (Bernardo Ostrowski, IFCN coordinator for Argentina, personal communication). The Santa Fe and Córdoba provinces produce 69% of the Argentine milk (Gutman et al., 2003). It is important to mention that this model does not represent the Argentine statistical average. The model that the panel of experts of the IFCN defined as being most similar to the statistical average for Argentina is a farm of 150 cows on 250 hectares (Bernardo Ostrowsky, personal communication). These characteristics of the average Argentine farm are similar to the results of a survey done by Gambuzzi et al. (2003) of 530 farms in the most important dairy farming regions of Argentina, where they found an average herd of 174 cows on a farm of 271 hectares. Consequently, this 350 cows, 450 hectare Argentine model has approximately twice the number of cows and 1.7 times the number of hectares compared to the statistically average farm (see Table 5).

Table 5: Differences in Size Between IFCN Models and Statistical Average for Argentina

	AR-350	Average Farm IFCN	Average Farm Gambuzzi et al.
Number of Cows	350	150	174
Hectares	450	250	271

The following table presents the main descriptive data of these two typical farm models:

---

[16] The 250 cows was not analysed in the present study, because that model appeared in IFCN (2002) and then was discontinued.

Table 6: Typical Farm Models for New Zealand and Argentina
(IFCN, 2002)

Farm		NZ-239	AR-350
Cows at the beginning (1)	no.	239	350
Live weight (2)	kg	437	500
Fat and protein content of milk	%	4.85% / 3.63%	3.5% / 3.2%
Milk yield per cow (3)	kg ECM	4,174	5,284
Land Use Dairy Enterprise			
Land use	Ha	104	450
Share of grassland	%	100%	85%
Share of maize silage	%	0%	15%
Share of grain and other	%	0%	0%
Milk produced per ha	t ECM / ha	9.6	4.1
Stocking rate (4)	LU / ha	2.7	1.2
Labour			
Dedicated to dairy (5)	Labour units	1.9	6
Share of family labour	% of total labour	67%	8%
Herd Management			
Seasonality	yes / no	yes	no
Calving season	Months	July - Aug	March - Dec
Age of first calving	Months	24	28
Intercalving period	Days	365	395
Dry cows (6)	%	17%	25%
Feeding			
Feeding system	Description	feed wagon	concentrate in parlour
Feed source winter (7)	Description	grazing off farm	grazing + silage
Feed source summer	Description	grazing	grazing
Grazing Season	duration in days	365	365
Concentrates use (8)	t DM total	19	753

Notes for Table 6:

1) Average number of dairy cows (dry and lactating) per season
2) Average live weights of cull cows.
3) Energy Converted Milk (ECM) with 4% milk-fat and 3.3% milk-protein[17].
4) Livestock Unit (LU), 1 LU is equal to 1 adult cow. The Stocking Rate includes the young stock and the calves raised on the farm.
5) One labour unit was assumed to work approximately 2,440 hours in New Zealand and between 2,555 and 2,700 hours in Argentina.
6) Proportion of average dry cows for the season.
7) In New Zealand it is common to send the young stock and some dry cows to graze off the farm. In Argentina it is normal to have annual winter pastures (oats and rye grass) because the perennial pastures have very low growth rates during the winter (IFCN, 2002).
8) Concentrate is any grain, soybeans meal, compound feed and all other comparable feedstuffs with high energy or protein content and dry matter > 85% (in tonnes).

---

[17] Formula used for adjustment: ECM = (milk production in litres * ((0.383 * % fat + 0.242 * % protein + 0.7832)/3.1138)

4.2.2 Contrast of NZ-239 and AR-350

The data for the following analysis was presented in Table 3. These two farm models (NZ-239 and AR-350) were defined as typical of the farms that contribute the biggest proportion of the total milk for each country.

Farm size: In comparison to the New Zealand dairy farm model, the Argentine dairy farm model is bigger in hectares (3 times more), in cow numbers (1.46 times more) and in total milk production (1.85 times more).

Cows' size and milk yield: The cows of the Argentine farm are heavier (1.13 times more) and produce more (1.21 times more) per cow than those of the New Zealand farm. Expressed as milk produced per kilogram of live weight, the value for the Argentine farm is slightly higher (1.09 more).

Calculations for NZ-239: 4174 kilograms ECM / 437 kilograms LW = 9.55 kilograms ECM/kilograms LW.

Calculations for AR-350: 5284 kilograms ECM / 500 kilograms LW = 10.57 kilograms ECM/kilograms LW

Land use: All the land used by the dairy enterprise in the New Zealand farm is utilised as grazing area, consequently all the concentrates (very little used) are brought in. In contrast, 15% of the land used in the Argentine farm is utilised for maize silage, grain or other crops.

Stocking rate: The New Zealand typical farm supports 2.7 livestock units (LU) per hectare. In contrast, the Argentine typical dairy farm supports slightly more than 1 LU. The differences between the cows must also be addressed; therefore if the stocking rate is calculated as kilogram of live weight per hectare, the New Zealand farm supports 1.97 times more livestock than the Argentine.

Calculations for NZ-239: 2.7 LU/ha x 437 kilograms LW/LU = 1180 kilograms LW/ha.

Calculations for AR-350: 1.2 LU/ha x 500 kilograms LW/LU = 600 kilograms LW/ha.

Herd management: The main difference between the management of the two herds is related to reproduction. The New Zealand average farm has a clearly defined calving season from July to August. In contrast, the cows of the Argentine farm calve from March to December. The average age of cows at first calving is 24 months on the New Zealand farm, and 28 months on the Argentine farm. The inter-calving periods are 365 days and 395 days respectively. The period in which the cows are not milked (dry period) is longer on the New Zealand farm (15 weeks) than the Argentine (7 – 10 weeks). The cows are milked for nearly two more months per lactation on the Argentine typical farm.

Feeding: Both farms graze their cows all year round. During the 2002 season, the New Zealand farm model bought some maize silage and the Argentine farm produced its own maize silage and bought some concentrates. The total requirements for NZ-239 were 1,143 tonnes of DM, and for AR-350 were 2,143 tonnes of DM.

In the New Zealand farm, 19 tonnes of concentrates, 40 tonnes of grass silage, and 26 tonnes of maize silage were bought (1.7%, 3.5% and 2.3% of total requirements, respectively). From the run-off 125 tonnes of grass were utilized for the young stock and dry cows (10.9% of total requirements). Therefore on the New Zealand farm 96% (100% – 1.7% - 2.3%) of the requirements were covered by pasture; and 85% (100% - 10.9 – 1.7% - 2.3%) of the total requirements were covered with homegrown feed.

In the Argentine farm, 469 tonnes of concentrates were bought (22% of total requirements). Homegrown feed was 225 tonnes maize silage, 1104 tonnes of grass, 265 tonnes of maize grain, and 71 tonnes of hay (11%, 52%, 12%, and 3% of total requirements respectively). Therefore in Argentina 52% (100%- 22% - 11% - 12% - 3%) of the requirements were covered by pasture; and 78% (100% - 22%) of the total requirements were covered with homegrown feed.

4.2.3 Four More Typical Farms

Four typical farm models from the IFCN are added to those previously described in section 4.2.1, two from New Zealand and two from Argentina. These farms are added in order to show some of the variability of typical dairy systems within each country.

New Zealand Typical Models

NZ-447: Represents a typical farm in the Southland region. This region has been the fastest growing region in total milk production in the last 10 seasons, and contributed 8% of the total production of the country in the 2002-2003 season (calculations based on LIC, 2002/03). Flat good quality land, temperate climate with colder winters, and a rainfall of 1,046 mm per annum characterize this region. The farm is a family business that supplies milk in a seasonal way; bought-in grass silage accounts for 5% and off-farm grazing accounts for 16% of the total feed requirements of the farm (IFCN, 2002).

NZ-835: Represents a typical farm in the South Canterbury region. The Canterbury region (the whole region, North and South together) is the second fastest growing region in milk production in the last 10 seasons, and contributed 11% of the total production of the country in the 2002-2003 season (calculations based on LIC, 2002/03). This region is characterized by flat good quality pastureland; temperate climate with low rainfall and cold winters, and is reliant on irrigation for pasture production. This typical farm is a family or equity partnership business that supplies milk in a seasonal way; bought in grass silage accounts for 20% and off-farm grazing accounts for 10% of the total feed requirements of the farm. This farm represents the biggest of New Zealand farms that are undergoing an expansion to 1,000 cows (IFCN, 2002).

Argentine Typical Models

The three Argentine typical farms are situated in the humid pampas, between 31° and 35° latitude, where over 90% of Argentine agricultural products are grown. Average temperature is 16.1C ° to 18.3C°, rainfall between 750 and 1,000 mm per annum, and 5 to 25 days with frosts. The main perennial pastures are lucerne (alfalfa), fescue and cocksfoot; the most common annual winter pastures are oats and rye grass. The dairy area coincides with the grain production area; this has two main consequences: on the one hand, grain is available at relatively competitive prices for milk production, and on the other hand, farmers may increase the area of land used for crop production by taking out an area from the dairy enterprise, when milk prices are low but crop prices are high (IFCN, 2002).

AR-150: Is a family farm that represents the Argentine average dairy farm in size and productivity. It would be theoretically situated in the northeast of the Córdoba province. Homegrown maize silage, maize grain and hay account for approximately 24%, 7% and

8% of the total feed requirements of the farm respectively. Bought-in concentrates account for 2% of total feed requirements of the farm. The balance (59% of total feed requirements of the farm) is from grazing lucerne (42% of total) and annual winter pasture (17% of total) (IFCN, 2002).

AR-1400: This model is based on data of a group of farmers from the west of Buenos Aires province. It is situated within the top 10% farms in size and within the top 25% in management level. The Buenos Aires province contributes 27% of the Argentine total milk production, and the largest farms are situated to the west of this province (Gambuzzi et al., 2003). Homegrown feed accounts for 96% of the feed requirements of the farm: Maize grain accounts for 8%, maize silage 17%, grazing perennial pasture (mix of fescue, cocksfoot, white clover and lucerne) accounts for 49%, and winter and summer crops 12%, soybean for grazing accounts for 9.9%, and hay 0.1% of total feed requirements of the farm. Bought-in concentrates account for 4% of total feed requirements of the farm (IFCN, 2002).

4.3 The IFCN Method for Dairy Farms

+ Total Returns (1) =
+ Milk returns
+ Non-milk returns =
+ Sales of cull cows
+ Sales of surplus heifers
+ Sales of calves
+ Interest on savings

- Total Expenses (1) =
+ Cash expenses
+ Paid wages (gross salary + insurance, taxes, etc)
+ Paid land rent
+ Paid interest on liabilities

= Net Cash Farm Income

+ Non cash adjustments =
- Depreciation (2)
+/- Change in livestock inventory (3)
+/- Capital gains / losses (4)
+/- VAT balance (5)

= Farm Income

- Opportunity costs =
+ Calculated interest on owned capital (6)
+ Calculated rent on owned land (7)
+ Calculated cost for unpaid family labour (8)

= Entrepreneur's Profit

Notes for the IFCN Method:

1) All cost components and returns are stated without value added tax (VAT), called GST in New Zealand.
2) Machinery and buildings were depreciated using a straight-line schedule on purchase prices with a residual value of zero.
3) Change in livestock inventory, is the difference in numbers between the beginning and the end of the seasons; it assumes that the value of livestock does not change during the season.
4) Capital gains / losses include the change in numbers and value of buildings and machinery. It does not include changes in land values; it assumes that the value of land does not change during the season.
5) Value Added Tax (VAT) balance.
6) Capital is defined as owned assets, without land and quota (calculation: assets for buildings, machinery, livestock and other), plus circulating capital (10% of all dairy related variable expenses), a real interest rate of 3% was used in all countries.
7) Regional rent prices provided by the farmers were used for owned land. In those countries with limited rental markets (like NZ), the land market value was capitalised at 4.5% annual interest to obtain a theoretical rent price.
8) For unpaid family labour, the average wage rate per hour for a qualified full-time worker in the respective region is used.

Another categorization of costs is used by the IFCN. This way of analysing the Total Costs is by "costs component", as described below:

Total Costs =
+ *Costs as Means of Production =*
+ Animal purchases
+ Feed (purchase feed, fertiliser, seed, pesticides)
+ Machinery (maintenance, depreciation, contractor)
+ Fuel, energy, lubricants, water
+ Buildings (maintenance, depreciation)
+ Vet & medicine, insemination
+ Insurance and taxes
+ Other inputs

+ *Capital Costs =*
+ Paid interest on liabilities
+ Calculated interest on capital

+ *Land Costs =*
+ Paid land rent
+ Calculated rent on owned land

+ *Labour Costs =*
+ Paid wages (gross salary + social fees)
+ Calculated cost for unpaid family labour

4.4 Performance of Typical Farms

Before starting the analysis it is important to mention that the farm models have
different management levels. The three New Zealand farms have management levels
that can be considered average for the country. However only the smallest Argentine
model (AR-150) has average management levels, AR-350 and AR-1400 are probably
within the top 25% in management levels. The panel of experts of the IFCN for each
country assessed the management levels. The differences in the management levels in
the Argentine (with the AR-150 having lower returns than the AR-350 and the AR-
1400) and the similarities in the New Zealand farms can be seen in indicators such as
the Return on Investment and the Operating Profit Margin (see Figures 14 and 15, and
Figures 16 and 17 further in the present chapter).

4.4.1 Milk Prices, Profits and Returns on Investment

The season that ended in 2002 is difficult to analyse for Argentina because on January
8th, 2002 there was a big devaluation of the Argentine currency that generated
uncertainty about costs and prices. Previous to the devaluation the exchange rate was 1
AR-$ for each US-$, whereas the first six months after the devaluation it was 2.6 AR-$
for each US-$ and 3.6 AR-$ for each US-$, for the last 6 months. The average for the
whole year was 3.1 AR-$ for each US-$ (*FXHistory: Historical currency exchange
rates*, 2005). Most of the costs started to increase due to inflation which was 118% for
that year (calculations based on Wholesale Price Index INDEC, 2005).

The 2003 season was more stable in Argentina, the exchange rate stayed constant at
nearly 3 AR-$ for each US-$ (*FXHistory: Historical currency exchange rates*, 2005)
and inflation was nearly 2% (calculations based on Wholesale Price Index INDEC,
2005).

Figure 10: Milk Price (season 2002)

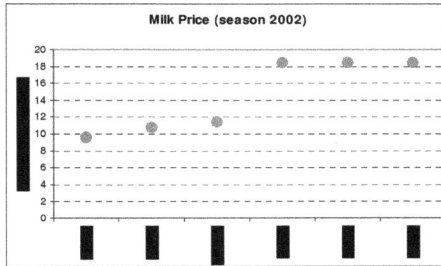

Figure 11: Milk Price (season 2003)

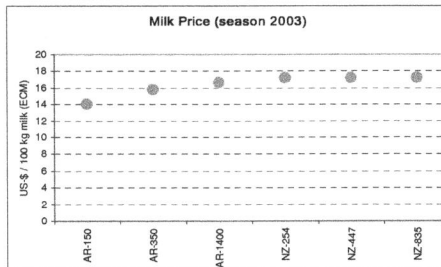

Milk prices are expressed per 100 kilograms of ECM (Energy Converted Milk; see section 3.2).

In the 2002 season the average price of milk for the typical New Zealand farms was US-$ 18 per 100 kilograms of milk (ECM) and for the typical Argentine farms US-$ 11. For New Zealand, this price was higher than in any of the previous 10 years (see section 3.1.3). For Argentina -despite the fact that this was the third best price in the previous 10 years (see section 3.2.3)- it was a difficult season in which both milk price and costs increased (in AR-$), under the influence of the devaluation of the Argentine currency.

In 2003, the milk price in New Zealand stayed relatively constant in US-$ but decreased in NZ-$ due to changes in the exchange rates. The 2003 season milk price was lower than in any of in the previous 10 years in NZ-$ for New Zealand farmers. By contrast, in Argentina the milk price increased remarkably in US-$ and in AR-$ in 2003, and was higher than in any (in AR-$) of the previous 10 years.

In conclusion, it can be said that 2002 was a wonderful season for New Zealand farmers and not very beneficial for Argentine farmers and the season 2003 was the opposite. This is more clearly seen in the following figures in which the profits and the return on investment are analysed.

Figure 12: Returns and Profits (season 2002)

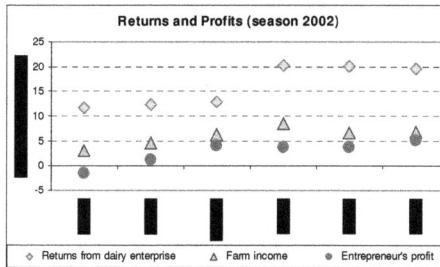

Figure 13: Returns and Profits (season 2003)

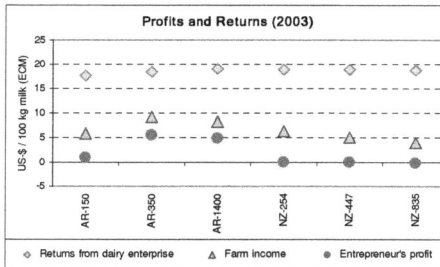

The Entrepreneur's Profit (calculated as Total Returns – Total expenses + Non Cash Adjustments – Opportunity Costs, see section 3.2) measures the economic sustainability of the business in the long run (IFCN, 2002). For the 2002 season, the New Zealand models showed a higher economic sustainability than two of the Argentine models. For 2003 the Argentine models had higher Entrepreneur's Profit.

Note that in the 2003 season, despite the fact that New Zealand and Argentine models had similar returns, the Farm Income and Entrepreneur's Profit were higher for Argentine models. This could be due to higher costs as means of production, by higher paid interest on liabilities because of the higher debt load of New Zealand farms, and other causes that are analysed further in this section.

Figure 14: Return on Investment (season 2002)

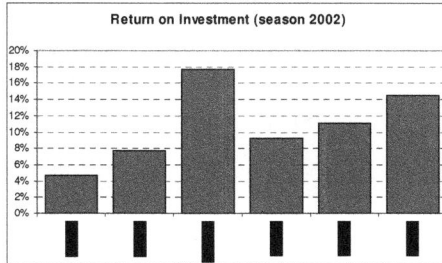

Return on Investment (season 2002)

Figure 15: Return on Investment (season 2003)

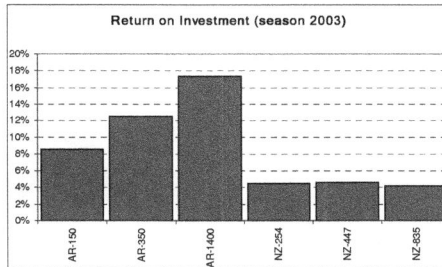

Return on Investment (season 2003)

The Return on Investment (ROI) is calculated by dividing the Operating Profit by the market value of all the assets invested in the business. The ROI is a factor of returns over assets invested. It is a relevant indicator for people who want to invest in dairy farming.

In 2002 the Argentine farm models had a ROI ranging from 3.3% to nearly 20%. The New Zealand farm models had a ROI ranging from more than 9% to nearly 15%.

In 2003 Argentine farms had a ROI ranging from 8.5% to 17.4%, and New Zealand farms from 4.2% to 4.5%. In 2003 asset values (specially land) increased in New Zealand, particularly in the South Island (where NZ-447 and NZ-835 are situated) and the returns decreased due to lower milk prices; hence ROI decreased remarkably in New Zealand from 2002 to 2003.

For all farm models, the ROI was higher than the 3% rate used as the opportunity cost of own capital, in both years (see section 4.3).

It is important to bear in mind the differences in the management levels of farm models. Only AR-150 has average management levels for Argentina, therefore in order to compare the return on investment of the farm models of the two countries this farm should strictly be compared with the New Zealand farms, all of which have average management levels. The AR-350 and AR-1400 have higher than average management levels.

4.4.2 Operating Profit Margin and Management Levels

The Operating Profit Margin is the ratio between the Operating Profit and the Total Returns of the business. The Operating Profit Margin is an indicator of the operating efficiency of the dairy farms; it indicates how well the farms have turned income into profit. The Operating Profit, also called Economic Farm Surplus (EFS) in New Zealand, is calculated in the following way:

Operating Profit =
> \+ Farm Income (see 2.3.4.3)
> \+ Paid land rent
> \+ Paid interest on liabilities
> - Calculated cost for unpaid family labour (opportunity cost)

The Operating Profit makes it possible to compare farms that operate on leased land with farms that operate on owned land. It also makes it possible to compare farms that are financed in different ways, and farms that have different proportions of unpaid family labour. It can also be used as an indicator of what financial commitments (how much interest and principal payments) a farm can carry (IFCN, 2002).

Figure 16: Operating Profit Margin (season 2002)

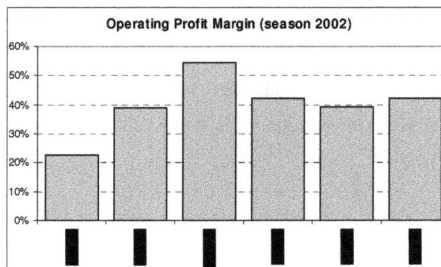

Figure 17: Operating Profit Margin (season 2003)

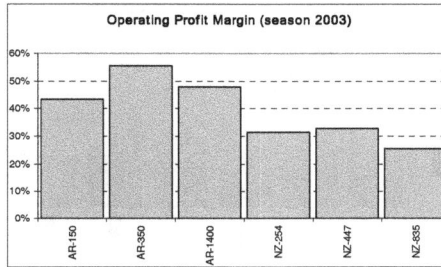

As was previously mentioned, the Operating Profit Margin gives an idea of the management levels (or operating efficiency) of the farms. Therefore, the fact that the two bigger Argentine farms have higher Operating Profit margins in both seasons than the smallest farm (see Figures 16 and 17) supports the IFCN experts who defined these as farms in the top 25% of farms for management levels.

For comparative purposes, the two farms that are more similar to the statistical average are AR-150 and NZ-239/254. In 2002, the Operating Profit Margin was 23% for the Argentine farm (AR-150) and was 42% for the New Zealand farm (NZ-239). In 2003 the Operating Profit Margin was 44% for the Argentine farm and was 31% for the New Zealand farm.

4.4.3   Milk Production Costs

After the devaluation of the AR-$ at the beginning of the 2002 season, the milk production costs, in US-$, of the Argentine typical dairy farms (IFCN) have decreased. This decrease in production costs has increased the possibility that the Argentine farm models can produce milk at lower costs than New Zealand farm models (see Figures 18 and 19).

Figure 18: Total Costs and Returns (season 2002)

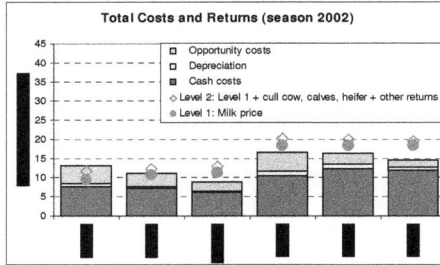

Figure 19: Total Costs and Returns (season 2003)

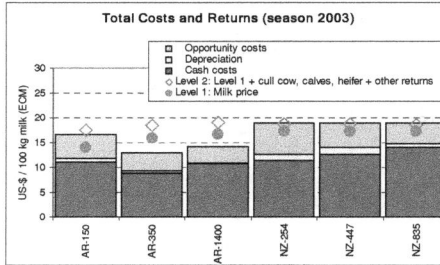

In 2002, Total Costs per 100 kilograms is lower for the three Argentine farms (US-$ 14) than for the three New Zealand farms (US-$ 19). In 2003 the average Total Costs were US-$ 19 and US-$ 24 respectively. The analysis of the costs follows in the next section.

4.4.4 Costs of Milk Production Only, and Non-milk Returns

The Costs of Milk Production Only is a useful indicator to compare the Total Costs that a dairy farm incurs to produce 100 kilograms of milk, with the price received for this milk. In order to calculate this indicator, the Non-milk Returns (returns from sales of cull cows, surplus heifers, calves, and other returns) are deducted from the Total Costs. This indicator allows the comparison of the costs of milk production of two farms in which the Non-milk Returns are different. For example, if two farms have the same Total Costs per 100 kilograms of milk, the farm with higher Non-milk Returns is able to produce milk at lower Costs of Milk Production Only. Essentially the Non-milk returns are treated as a sub-product of milk production.

Figure 20: Costs of Milk Production Only (season 2002)

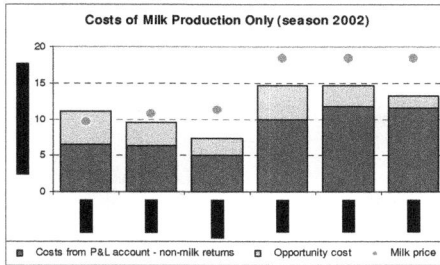

Figure 21: Costs of Milk Production Only (season 2003)

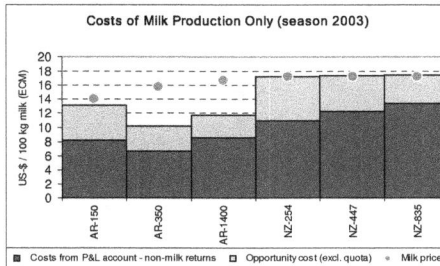

Figures 20 and 21 (Costs of Milk Production Only) show similar trends as those shown in Figures 18 and 19 (Total Costs), comparing the Argentine and the New Zealand farm models. The Costs of Milk Production Only is lower for the three Argentine farms than for the three New Zealand farms for both seasons. In 2002 the average Costs of Milk Production Only of the Argentine model farms was US-$ 11 and the average Costs of Milk Production Only of the New Zealand model farms was US-$ 15. In 2003 the average Costs of Milk Production Only were US-$ 12 and US-$ 17, respectively.

It is important to mention that before the devaluation of the Argentine currency (AR-$) in January 2001, New Zealand farms had lower Costs of Milk Production Only than the typical Argentine farms (IFCN, 2002). The actual exchange rate of the Argentine currency has been floating around AR-$ 3 for each US-$ since March 2003 with low intervention of the Argentine Central Bank, and is expected to continue in similar levels in 2005 (BCRA, 2005).

Figure 22: Non-milk Returns (season 2002)

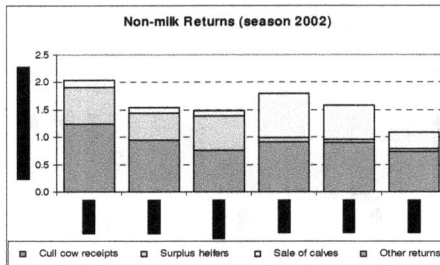

Figure 23: Non-milk Returns (season 2003)

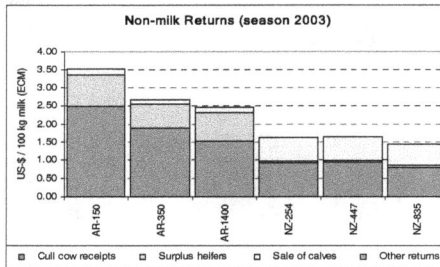

The Non-milk Returns per 100 kilograms of milk are the returns from sales of cull cows, the surplus heifers, and the calves.

In 2002, the total Non-milk Returns were similar for farms in both countries (US-$ 1.5 per 100 kilograms of milk on average for the three Argentine farms and US-$ 1.7 for New Zealand farms). Non-milk Returns were 14% of the Total Returns for the Argentine farms and 7% for the New Zealand farms (see Table 7).

In 2003 the Non-milk Returns were higher for the Argentine typical farms than for New Zealand farms (average of the three farms US-$ 2.9 for Argentina and US-$ 1.6 for New Zealand). Non-milk Returns were 16% of the Total Returns for the Argentine farms and 8% for the New Zealand farms (see Table 7).

As shown in Figures 22 and 23, the proportion of the Non-milk Returns that is derived from selling off surplus heifers is higher for Argentine farms than for New Zealand farms (see Table 7). This is partly caused by the fact that most Argentine farms raise all their heifer calves (researchers' experience in Argentine dairy farming). Additionally, these figures show that for New Zealand farms, the proportion of the Non-milk Returns that comprises the selling of calves is higher (see Table 7). This is because New Zealand farms sell not only their bull calves but also a proportion of their heifer calves. In addition in Argentina the prices received for bull calves are nearly insignificant because there is no market for them. In contrast, in New Zealand, there is an established market for bull calves also called "bobby calves" (NZ-Biosecurity, 1997).

Table 7: Non-milk Returns, as a Percentage of Total Returns, for New Zealand and Argentine Typical Farms

	Argentine Typical Farms		New Zealand Typical Farms	
	Season 2002	Season 2003	Season 2002	Season 2003
Cull cows	8.1%	10.7%	4.3%	4.6%
Surplus Heifers	4.8%	4.3%	0.3%	0.3%
Calves	0.9%	0.8%	2.9%	3.4%
Total Non-milk Returns	13.8%	15.8%	7.5%	8.3%

Table 7 also shows that the proportion of the returns from the sales of cull cows in Total Returns is higher for the Argentine model farms than for the New Zealand model farms.

As reflected in Figures 22 and 23, the returns from the sale of cull cows per kilogram of milk produced were higher in the Argentine farms than in the New Zealand farms (1.9 times higher in 2002 and 2.2 times higher in 2003). This is explained by the higher replacement rates of the Argentine farms and the bigger size of the Argentine cows. The replacement rate was, on average, 29% for the Argentine farms and 21% for the New Zealand typical farms. The live weight of the typical Argentine cows is between 500 and 550 kilograms and for the New Zealand typical cow is approximately 440 kilograms.

4.4.5 Costs components (IFCN)

In the following sections, the different cost components are analysed, as defined by the IFCN: Labour Costs, Land Costs, Capital Costs and "Costs as Means of Production" (see section 4.3).

Table 8: Cost Components, as Percentage of Total Costs, for New Zealand and Argentine Typical Farms

	Argentine Typical Farms		New Zealand Typical Farms	
	Season 2002	Season 2003	Season 2002	Season 2003
Labour Costs	19%	17%	18%	18%
Land Costs	25%	24%	13%	16%
Capital Costs	4%	5%	11%	11%
Costs as Means of Production	52%	54%	58%	53%
Total Costs	100%	100%	100%	100%

4.4.6 Labour

In this section, Labour Costs per 100 kilograms of milk, the Average Wages of the farms, and the Labour Productivity are analysed.

Figure 24: Labour Costs (season 2002)

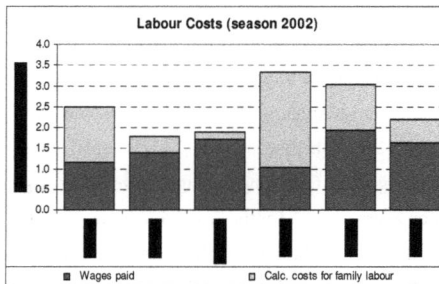

Figure 25: Labour Costs (season 2003)

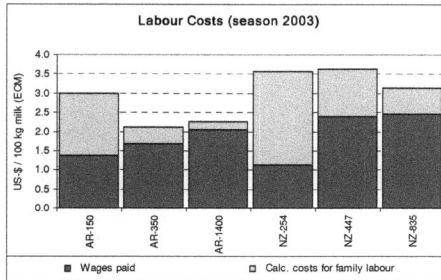

On New Zealand farms Labour Costs were, on average, 1.4 times higher per 100 kilograms milk than on the Argentine farms in the 2002 season and in the 2003 season.

Figure 26: Average Wages (season 2002)

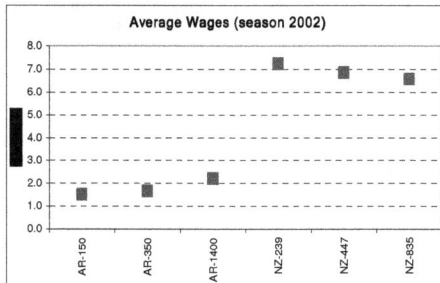

Figure 27: Average Wages (season 2003)

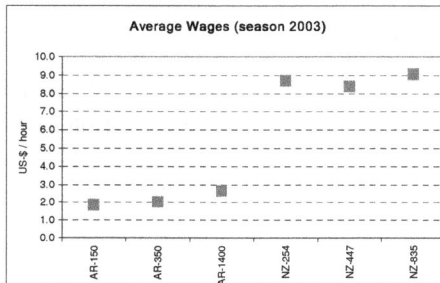

In 2002 Average Wages paid per hour of work on New Zealand farms were 2.8 times higher than on Argentine typical dairy farms. In 2003, they were 4 times higher.

Figure 28: Labour Productivity (season 2002)

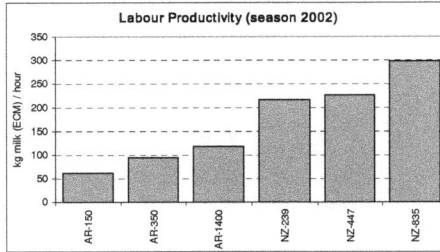

Figure 29: Labour Productivity (season 2003)

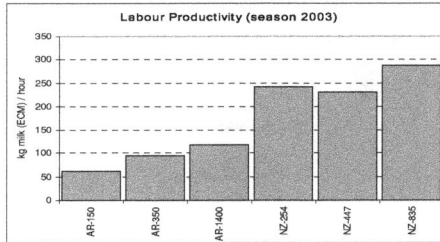

Labour Productivity is calculated by dividing the total kilograms of milk (ECM) produced by the total hours for feeding, forage production, young stock rearing, milking and manure handling (IFCN, 2002).

In the New Zealand farm models the Labour Productivity (calculated as kilograms of milk produced per hour of work) was 2.7 times higher than in Argentine farm models in 2002, and 2.8 times higher in 2003.

In conclusion, Labour Costs per kilogram of milk produced were higher in New Zealand than Argentina, despite the fact that Labour Productivity (in kilograms of milk per hour of work) was higher in New Zealand than in Argentina, because the difference on average Wages (US-$ per hour of work) were larger than the differences in Labour Productivity.

4.4.7 Land

Land comprised more the half of the total assets for New Zealand and Argentine farms. For the Argentine typical farms land was 77% of the total assets in 2002 and 81% in 2003. For the New Zealand farms land was 56% of the total assets in both seasons.

In this section the Land Costs per 100 kilograms of milk, the Market Value of Land, the Level of Land Rents and Land Productivity are analysed.

Figure 30: Land Costs (season 2002)

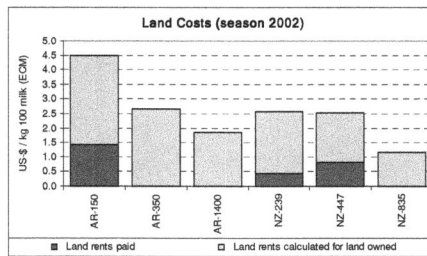

Figure 31: Land Costs (season 2003)

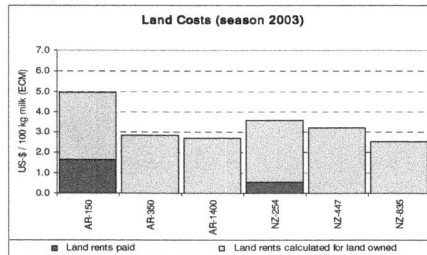

Total land costs are the sum of the real land rents paid to the opportunity value of land (see section 4.3). Land Costs are calculated by dividing the total land costs by the kilograms of milk produced.

In Argentine farm models Land Costs were, on average, 1.28 times higher per 100 kilograms milk than in the New Zealand farms in the 2002 season, and in the 2003 season 1.12 times higher. When comparing farms of similar management level, the

Land Costs of AR-150 were 1.7 and 1.4 times higher per kilogram of milk produced in 2002 and 2003 than those of NZ-239/254.

In Argentina it is more common to rent a piece of land than in New Zealand (researcher's experience in Argentine dairy farming sector). This cannot be seen in Figures 30 and 31 because only the AR-150 leased land, and because the run-offs (land separated from the farm that is used for grazing dry cows, rearing replacement heifers or planting crops) of New Zealand farms appeared as rented land.

Figure 32: Market Value of Land (season 2002)

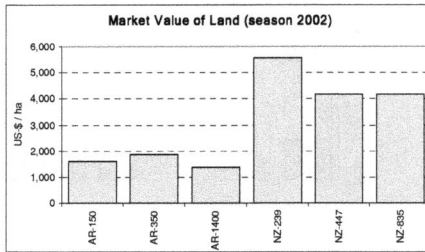

Figure 33: Market Value of Land (season 2003)

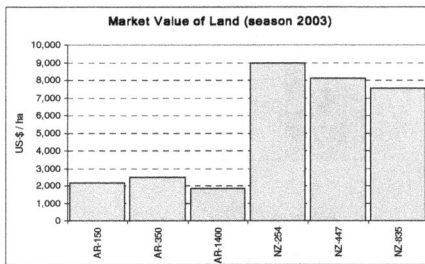

Average Market Value of Land in New Zealand farms was 3.75 times higher than in Argentine farms in both years.

Figure 34: Level of Land Rents (season 2002)

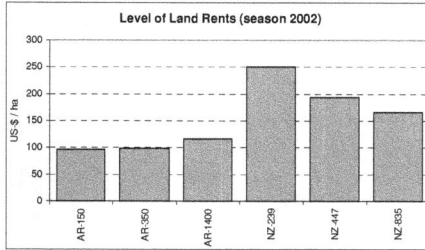

Figure 35: Level of Land Rents (season 2003)

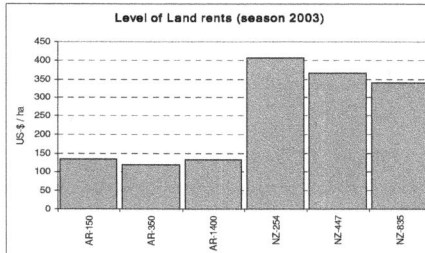

In 2002 the average Level of Land Rents in New Zealand farms was 2 times higher than in Argentine farms, and in 2003, 2.9 times higher. The Level of Land Rents was approximately 4.5% of the Market Value of Land in New Zealand and approximately 6.5% in Argentina.

Figure 36: Land Productivity (season 2002)

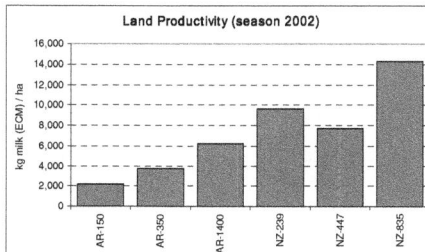

Figure 37: Land Productivity (season 2003)

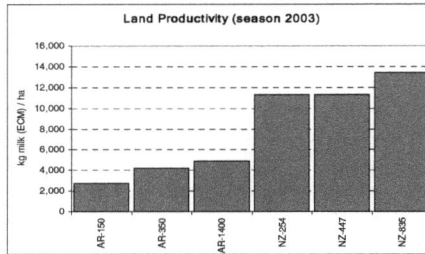

Land Productivity is calculated by dividing the total milk produced in kilograms of milk (ECM) by the hectares utilized by the dairy enterprise. In the Argentine models most of the supplements are produced on the same farm. Those hectares utilized for producing those supplements are included in the calculation of the Land Productivity; the only feeds that are not included are the concentrates that were bought. In the New Zealand farms, the run-off hectares are not included to calculate the Land Productivity consequently the grass grown in the run-off is feed that was brought in.

Should be mentioned that the amount of feed brought in from outside the hectares considered has an impact in land productivity:

- Typical New Zealand farms: NZ-239 brought in 15% of total feeds, NZ-447 brought in 21% of total feeds, and NZ-835 brought in 30% of total feeds (the proportions are approximate).

- Typical Argentine farms: AR-150 brought in 0.5% of total feeds, AR-350 brought in 22% of total feeds and AR-1400 brought in 10% of total feeds (the proportions are approximate).

In 2002 the average Land Productivity of New Zealand farm models was 2.6 times higher than in Argentine typical dairy farms, and in 2003, 3 times higher.

In conclusion, the Land Costs per kilogram of milk produced were higher in Argentina than in New Zealand, despite the fact that the Market Value of Land and the Level of Land Rents were lower in Argentina than in New Zealand. This is explained because the difference in Land Productivity (kilograms of milk per hectare) was even higher than the difference in Market Value of Land and Level of Land Rents.

4.4.8 Capital

Capital is another cost component. The IFCN analyses the capital factor separately from the land factor, consequently capital in the present context is the total assets without the land (buildings, machinery, livestock, co-operative shares, and others).

Figure 38: Capital Costs (land not included) (season 2002)

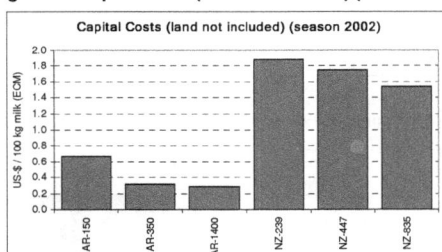

Figure 39: Capital Costs (land not included) (season 2003)

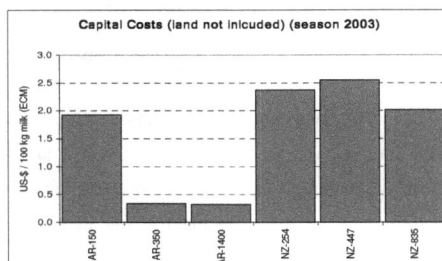

In order to calculate the Capital Costs (on an annual basis), the IFCN considers a real interest rate of 6% for borrowed funds, and a 3% real interest (as opportunity cost) for owner's capital (buildings, machinery, livestock and others).

In 2002 the average Capital Costs in New Zealand farms were 4 times higher than in Argentine typical dairy farms, and in 2003, 2.7 times higher. The Capital Costs of AR-150 changed dramatically from 2002 to 2003 because the farm had financial problems during 2002 and incurred in a large short-term loan that was paid during 2003. The following figures (Capital Structure, Capital Input per cow, Milk Yield per cow and Capital Productivity) help to understand why Argentine farms had less Capital Costs.

Figure 40: Capital Structure of the Typical Farms (season 2002)

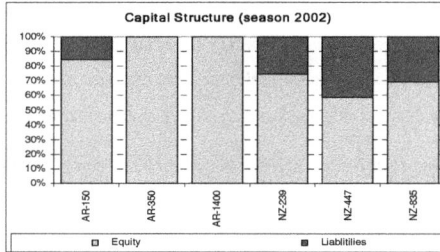

Capital Structure (season 2002)

Figure 41: Capital Structure of the Typical Farms (season 2003)

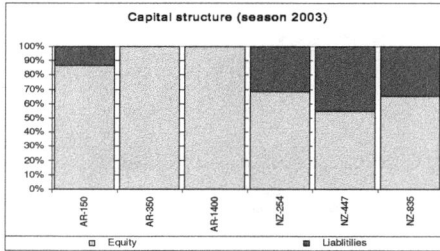

Capital structure (season 2003)

New Zealand typical farms had, on average, 33% of liabilities in 2002 and 37% in 2003. Argentine typical farms had, on average, 5% of liabilities in 2002 and 4% in 2003. As shown in Figures 40 and 41, two of the three typical Argentine dairy farms do not have debts. This is because, among other things, the interest rates are usually very high in Argentina and because loans are difficult to obtain.

Figure 42: Capital Input in Buildings and Machinery per Cow (season 2002)

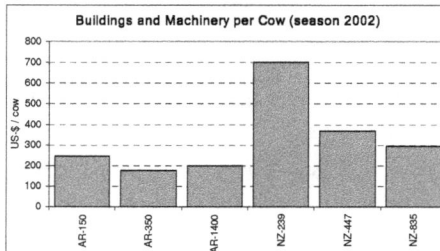

Buildings and Machinery per Cow (season 2002)

Figure 43: Capital Input in Buildings and Machinery per Cow
(season 2003)

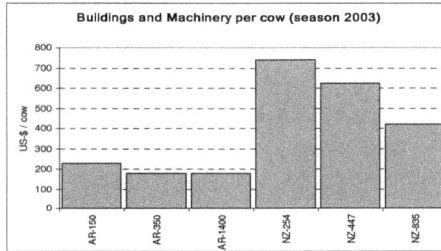

Capital Input in Buildings and Machinery is calculated by dividing the total assets of buildings and machinery of each farm divided by the number of cows.

In 2002 the average Capital Input in Buildings and Machinery per cow in New Zealand typical farms were 2.2 times higher than in Argentine typical dairy farms, and in 2003, it was 3.1 times higher.

Figure 44: Price per Heifer Sold (season 2002)

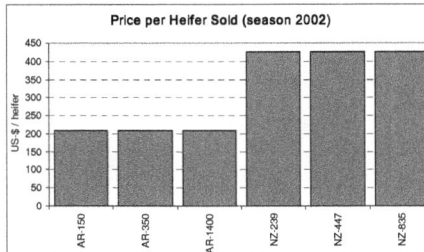

Figure 45: Price per Heifer Sold (season 2003)

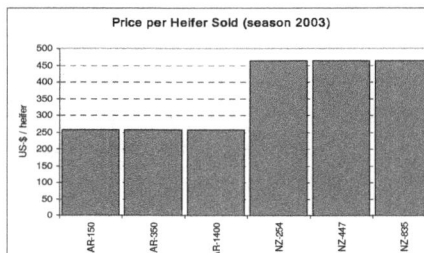

In 2002 the average Price per Heifer Sold (pregnant heifers 20 months old) in New Zealand typical farms was 2 times higher than in Argentine typical dairy farms, and in 2003 1.8 times higher.

This difference in Price of Heifers Sold between countries is similar for the other categories of livestock; which explains why the total capital input in livestock divided by the number of cows (Capital Input in Livestock per cow) was 2.65 times higher in New Zealand model farms than in Argentina model farms for both seasons. This is caused by lower livestock prices in Argentina than in New Zealand.

Figure 46: Capital Productivity (land not included) (season 2002)

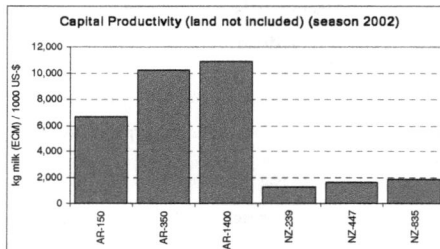

Figure 47: Capital Productivity (land not included) (season 2003)

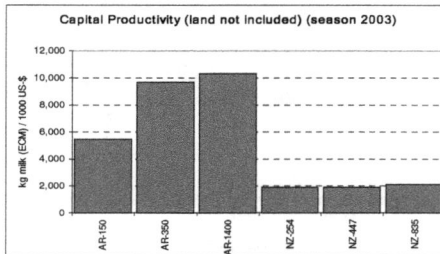

Capital Productivity is calculated by dividing the total milk produced by the capital invested in buildings, machinery, livestock, and co-operative shares. In 2002 capital productivity in Argentine typical farms was 5.8 times higher than in New Zealand typical dairy farms, and in 2003, it was 4.3 times higher.

This difference is caused by the fact that the Argentine typical farms produced more milk per cow (see Figures 48 and 49) with less capital invested per cow.

Figure 48: Milk Yield per Cow (season 2002)

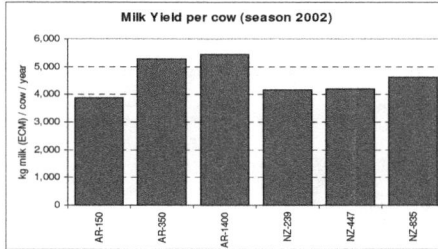

Figure 49: Milk Yield per Cow (season 2003)

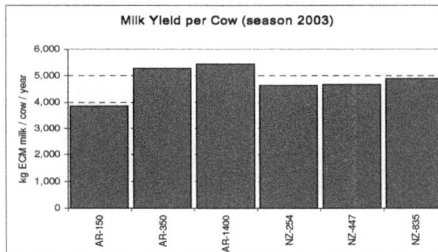

In 2002 the Milk Yield per cow in Argentine typical farms was 1.12 times higher than in New Zealand typical dairy farms, and in 2003 1.03 times higher.

In conclusion, Capital Costs per kilogram of milk produced were higher in New Zealand than in Argentina mainly due to three causes: firstly because the New Zealand dairy farms have more than 6 times the amount of debt. Secondly, because the New Zealand dairy farms invested more in buildings, machinery and livestock per cow than the Argentine farms. And thirdly, because typical New Zealand farms need to own co-operative shares in order to sell their milk to the dairy company.

4.4.9 Costs as Means of Production

The last cost component defined by the IFCN is the Costs as Means of Production (see section 4.3). The means of production are the following: animal purchases, feed expenses (purchase feed, fertiliser, seed, pesticides), machinery expenses (maintenance, depreciation, contractor), fuel, energy, lubricants and water expenses, buildings expenses (maintenance, depreciation), veterinarian and medicine expenses, insemination expenses, insurance and taxes, and other inputs.

Figure 50: Costs as Means of Production (season 2002)

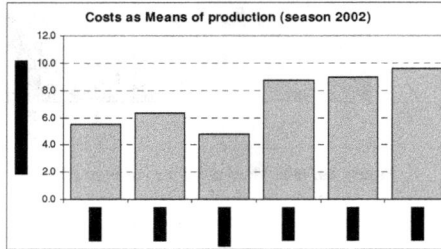

Figure 51: Costs as Means of Production (season 2003)

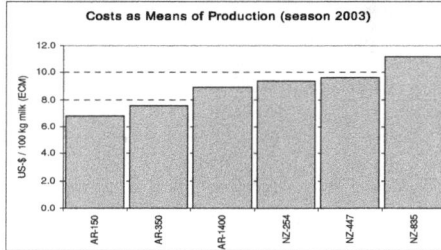

In 2002 the average Costs as Means of Production in New Zealand farms were 1.6 times higher than in Argentine typical dairy farms, and in 2003, they were 1.3 times higher. The fact that the difference decreased from 2002 to 2003 could be related by the increase of inputs in Argentine farms due to the increase in the milk price.

4.5 Chapter Summary

- The average New Zealand farm has 285 cows and 111 hectares.

- The average Argentine farm has 174 cows and 271 hectares.

- Argentine typical cows (IFCN) have a live weight between 500 and 550 kilograms. New Zealand typical cows weight approximately 440 kilograms.

- The typical farm models (IFCN) that contribute the biggest proportion of the total milk for each country are AR-350 and NZ-239.

- Cows in AR-350 and NZ-239 produce similar kilograms of milk (ECM) per kilogram of live weight.

- NZ-239 supports nearly 2 times more kilograms of live weight of livestock per hectare than AR-350.

- Typical Argentine farms have all year round calving. New Zealand typical farms have seasonal calving.

- In AR-350, 52% of the total feed requirements of all the livestock are covered by grass. In NZ-239, 96% of the total feed requirements of all the livestock are covered by grass.

- In AR-350, 78% of the total feed requirements of all the livestock are covered by homegrown feeds. In NZ-239, 85% of the total feed requirements of all the livestock are covered by homegrown feeds.

- The farm models defined by the IFCN as typical are AR-150, AR-350 and AR-1400. For New Zealand the typical farm models are NZ-239, NZ-447 and NZ-825.

- AR-150 and NZ-239 are the typical farms that are closer to the statistical dairy farm averages of their countries.

- AR-1400 represents a group of Argentine farms that are situated within the top 10% in size and within the top 25% in management.

- NZ-447 and NZ-835 are average farms from the two dairy farming regions that have been growing most rapidly in milk production in New Zealand during the last 10 seasons.

- New Zealand typical farm models had, on average, higher Costs for Milk Production Only per kilogram of milk (ECM) than Argentine typical farm models in 2002 and 2003 (1.6 times and 1.42 times higher, respectively).

- New Zealand typical farm models had, on average, higher Labour Costs per 100 kilograms of milk (ECM) than Argentine farm models in 2002 and 2003 (1.39 times and 1.41 times higher, respectively).

- New Zealand typical farm models had, on average, higher Capital Costs, per kilogram of milk (ECM) than Argentine typical farm models in 2002 and 2003 (2.1 times and 2.7 times higher respectively).

- New Zealand typical farm models had, on average, higher Costs as Means of Production, per kilogram of milk (ECM) than Argentine typical farm models in 2002 and 2003 (1.6 times and 1.3 times higher, respectively).

- Argentine typical farm models had, on average, higher Land Costs per kilogram of milk (ECM) than New Zealand typical dairy farm models in 2002 and 2003 (1.28 times and 1.12 times lower, respectively).

5   METHODOLOGY

5.1   Research Strategy

5.1.1 The Case Study

Several research strategies were analysed from the literature in order to find the most suitable strategy for the research question. Philliber, Schwab and Sloss (1980) suggested three main strategies for social research: case studies, surveys and experiments. Bouma (2000) incorporated two more research strategies: longitudinal studies and comparison studies. Another author who specializes in social research and more specifically in case studies (2002; Yin, 2002) also analysed the three most common (the experiment, the survey and the case study) and added two more to the five suggested by Bouma (2000), these are the archival analysis and the history.

From the seven research strategies named, the case study was chosen as the most appropriate for this research. The definition of the case study research strategy and the reasons for choosing it are explained in the following paragraphs.

There are very different definitions of the case study as a research strategy. Philliber et al. (1980, page 63) stated, "The distinguishing characteristic of the case study is that there is a sample size of one" and that "While many data-collection techniques may be employed in any design, multiple measures are generally the rule rather than the exception in case study research."

Bouma (2000, page 90) also stated that the main characteristic of the case study "is that it focuses on a single case" and that is the research strategy more suitable for exploratory research.

Other authors (Chetty, 1996; Eisenhardt, 1989, 1991; Rowley, 2002; Yin, 2002) defined the case study as a more powerful research strategy because they consider the possibility of multiple case studies and to build theories from case study research. They suggested that this research strategy could be used for many cases and for descriptive and explanatory research as a contrast to only exploratory research of only one case.

A case study can be defined as the following:

> "A case study is an empirical inquiry that investigates a contemporary phenomenon within its real-life context, especially when the boundaries between phenomenon and context are not clearly evident." (Yin, 2002, p.13)

Philliber et al. (1980) and Bouma (2000) agree with this definition of a Case Study. Both authors stated that the case study investigates a phenomenon within its real context and focused more in the present than in the past. Considering these two aspects, an experiment would be an example of a research strategy in which the researcher tries to have control over the context; and a history an example of a research strategy in which the researcher is more interested in the past than in a contemporary phenomenon.

It can be added to the definition of the case study that it usually employs many data collection techniques (Philliber et al., 1980). Therefore the data collected can be a mix of qualitative and quantitative data. Is also important to mention that the case study is the preferred strategy for exploratory research (Bouma, 2000), even though it can be used also for descriptive and explanatory research (Chetty, 1996; Eisenhardt, 1989, 1991; Rowley, 2002; Yin, 2002).

In this research project the phenomena to be studied is the impact of adopting New Zealand innovations to Argentine farms, the impact of this and the possibilities and constraints of diffusing the innovations through the Argentine production sector. Each farm is a case and it is impossible to separate the farms from their environment. Consequently, according to its definition, the case study is the most appropriate strategy for this research project.

5.1.2 Descriptive Research

In the continuum from exploration to explanation passing through description, this research started exploring and is mainly descriptive. It was not the objective of the present study to proved cause-effect relationships. However some associations between variables were found.

At the moment when the proposal for this research was elaborated no other study on the topic of this research could be found. Nor was there any trustworthy information about the following points:

1) There was not a list of New Zealand ideas, practices and technologies that were potentially useful for Argentine dairy farmers.

2) The proportion, number or identity of the Argentine dairy farmers that were adopting New Zealand ideas was unknown (but the existence of a group of farmers that were adopting New Zealand ideas was certain).

3) It was unknown which were the New Zealand ideas that had been adopted by the farmers in contact with the New Zealand ideas.

4) No study of the process of adoption of innovations (of any origin or kind) by Argentine dairy farms could be found.

5) As a consequence of all this it can be considered that the topic was unexplored and that this was the first study to investigate the mentioned topic.

5.1.3 Case Study Designs

In the following section are described the different kinds of case study designs. The one that was chosen is then described in detail.

Table 9: Basic Types Designs for Case Studies

	Single-case designs	Multiple-case designs
Holistic **(Single unit of analysis)**	Type 1	Type 3 Literal and/or theoretical replication
Embedded **(Multiple unit of analysis)**	Type 2	Type 4 Literal and/or theoretical replication

(Source: Adapted from Yin, 2002, p.40)

There are four basic types of case studies designs (see Table 8). Within single and multiple case studies designs there are two different types, the holistic and the embedded. An embedded case study is one in which more than one unit is analysed, for example a case study of a farm business in which every aspect of the business is investigated. On the other hand, a holistic case study is when only one unit is analysed, as in a case study of a farm business in which only the attitude towards risk of the farmer is investigated (for more details refer to Yin, 2002, p.42-45).

The present study is an embedded study because three units of analysis were investigated for each case study. The first unit of analysis was to assess the adoption or rejection of a list of New Zealand innovations. The second unit of analysis is the reasons that support the decision to reject or to adopt the innovations. The third unit is to study the impact of adopting the New Zealand innovations on each case study farm.

Also Table 8 shows that within the multiple case study designs (types 3 and 4) literal and/or theoretical replication can be chosen. Before explaining the two kinds of replications that can be used, it is important to clarify why the term "replication" was used rather than "sampling".

The sampling logic is commonly used in surveys in which the researcher selects a group of cases that statistically represents the population that is to be studied (Jankowicz, 1991 cited in Ghauri & Gronhaug, 2001). The objective of this kind of research is to generalize conclusions to the population. In this kind of research the ultimate object of study is not the sample, it is the population itself that is studied through the sample.

In contrast the replication logic is used when a specific phenomenon is studied by itself, not as a part of a greater phenomenon. In a multiple case study each case is studied as an end in itself with the intention to understand certain aspects of the case study. In a case study design of multiple cases in each case the same phenomenon is studied. It is similar to a multiple experiments design in which the same phenomenon is studied in more than one experiment in similar conditions. The intention in both the multiple experiments and case studies designs is to repeat similar studies of the same phenomenon with the intention to validate the results by repetition (Yin, 2002).

Continuing with the explanation about the replication logic, it was mentioned before that the replication could be literal or theoretical. Literal replication is when the cases are carefully selected with the intention to predict similar results. When someone does a literal replication he or she tries to find cases that are similar in the aspects that affect the unit of analysis that is being studied. Is easy to understand with an example of an experiment, the replication of experiments are tried in very similar conditions to validate the previous experiments. Theoretical replication is when each case is selected to predict contrasting results but for predictable reasons. Consequently when a multiple case study with theoretical replication is designed cases are selected that are different, with the intention to study the same phenomenon in cases that are selected to be different. In the example of an experiment, the aim would be to do the same experiment in slightly different conditions to test if the same results are obtained (Yin, 2002, p. 47).

In this research project the case study farms were selected with the intention to replicate literally and theoretically. Farmers considered innovators (in respect to New Zealand ideas) were selected with the intention to find similar results and some contrasting farmers (considered more typical or conservative) were also selected with the objective to contrast their results with the others (more details of the case selection in the next section).

In conclusion, the research is a multiple case study design because seven farms were selected with different degrees of innovativeness with respect to New Zealand innovations. And the cases were embedded because multiple units of analysis were investigated in each case (see Figure 52).

Figure 52: Research Design, Multiple and Embedded Case Study

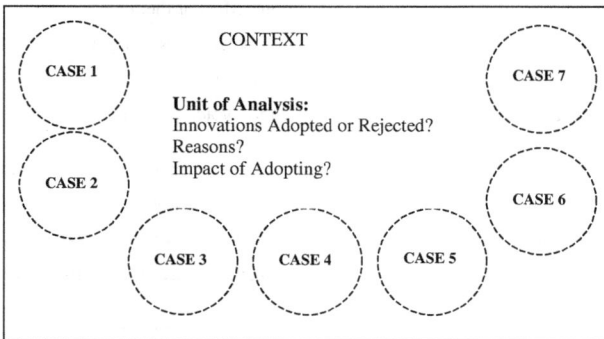

Figure 52 shows the main elements of the research study: the cases, the context and the unit of analysis. The dashed lines that define each case study show that the boundaries between the context and the cases are not clear. This is because the context and the cases are related. Farms and farmers are very influenced by the Argentine socio-economic environment and also by the changes that occur within the Argentine dairy sector and at the same time farmers are actively impacting on the Argentine dairy sector individually and as a group.

5.1.4 Cases Selection

For reasons of confidentiality numbers instead of real names identifies the farms and farmers in the present report, and specific details that could lead to accurate identification were omitted.

Farmers with knowledge of New Zealand systems were needed. It was essential to achieve the research objectives that the farmers that were to be chosen knew about the New Zealand dairy production systems and the New Zealand innovations prior to meeting the researcher.

The researcher established contact with one of these farmers, *Farmer 2*, who is renowned as an innovator within the dairy farm sector. The researcher, who is family related to *Farmer 2*, worked on one of *Farmer's 2* dairy farms for a year. Working on this farm the researcher became aware that there was a small group of Argentine farmers that were adopting New Zealand innovations. Most of them were interconnected because they shared information and experiences. They also applied systems that were different from typical Argentine farmers and consequently they were continually challenging and being challenged by more typical farmers.

Farmer 4 was recommended by *Farmer 2* as one of the first ones to adopt New Zealand innovations and as one of those most persuaded of the benefits of the New Zealand dairy farm principles. These two farms are very big in scale (more than 2,000 cows) and are probably representative of the biggest and best managed Argentine dairy farms.

These two farmers recommended *Farmer 6* as someone with comprehensive knowledge of the New Zealand dairy principles. This farmer has long experience in adopting the New Zealand dairy farm principles in several countries (for example, Nigeria, Jamaica and Nicaragua). This farm can be considered a family farm that is growing and being developed financed by its own profits. They started milking 6 years ago and have been increasing cow numbers up to the current 400 cows.

Farms 5 and *7* were recommended by *Farmer 2.* These two farmers had previously worked in association with *Farmer 2* and were at that moment working independently. They have experience in typical Argentine farms and also the experience of working in *Farm 2* where several New Zealand innovations have been adopted. Nowadays both of them are owners and directors of their own dairy businesses. Both farmers lease all the land and are probably exposed to a greater degree of risk than all the other farmers selected. *Farm 7* has currently 400 cows, is a family business in which the family provides part of the labour. In *Farm 5,* 1,000 cows were milked last season (03-04), and is also a family business. The fact that both lease all the land and that both farmers' families are very involved in the business obliges the farmers to be very cautious in the innovations that they adopt.

Farm 1 is a very big farm (6,500 cows) owned by a family that also owns several other farms and businesses. In this farm the CEO (Chief Executive Officer) was interviewed as the main decision taker of the dairy enterprise. This farm is situated in the more traditional and productive dairy area of Argentina. The 6,500 cows are milked in 18 farm dairies of different sizes. Historically *Farm 1* has been considered typical and with some degree of opinion leadership among other farmers in the area (people of the region interviewed). If the NZ innovations are proved to work in this farm it is possible that they will be spread to other dairy farmers.

Farmer 3 has a family relationship to *Farmer 2* and is in the same discussion group as *Farmer 4.* Therefore he is constantly in contact with the New Zealand innovations adopted in those two farms. *Farm 3* is also large scale (1900 cows). *Farmer 3* is considered to adopt new technologies and ideas only when they have proved to be beneficial (from interviews of *Farmer 2* and *Farmer 4).* This farm is among the top performers of the discussion group (data from the discussion group and interviews with *Farmer 2* and *Farmer 4).* Due to his prestige and relative acceptance of the social system norms *Farmer 3* can be considered to be an opinion leader within the farmers of the discussion group. Even though *Farmer 3* is considered sceptical about new technologies he has adopted some NZ innovations.

The researcher holds these farmers in high esteem; all of them are successful dairy farmers with great experience in the production sector, optimistic about the future of the dairy sector and actively compromised to its development.

5.1.5   Enhancing the Quality of the Research

Traditionally four criteria are used to examine the quality of any empirical social research. These are **construct validity**, **internal validity**, **external validity** and **reliability** (Kidder & Judd, 1986, p.26-29 cited in Yin, 2002):

> "**Construct validity:** establishing correct operational measures for the concepts being studied.
>
> **Internal validity:** establishing a causal relationship, whereby certain conditions are shown to lead to other conditions, as distinguished from spurious relationships.
>
> **External validity:** establishing the domain to which a study's findings can be generalized.
>
> **Reliability:** demonstrating that the operations of a study –such as the data collection procedures- can be repeated, with the same results."

To enhance the quality of this investigation some tactics that were suggested by Yin (2002) for different phases of a case study were used (see Table 10).

Table 10: Case Study Tactics for Four Design Tests

Tests	Case Study Tactic	Phase of Research in which Tactic Occurs
Construct validity	• Use multiple sources of evidence • Establish chain of evidence • Have key informants review draft case study report	Data collection Data collection Composition
Internal validity	• Do pattern-matching • Do explanation-building • Address rival explanations • Use logic models	Data analysis Data analysis Data analysis Data analysis
External validity	• Use theory in single-case studies • Use replication logic in multiple-case studies	Research design Research design
Reliability	• Use case study protocol • Develop case study database	Data collection Data collection

(Source: Yin, 2002, p.34)

Construct Validity

Construct validity addresses the question: *Are we measuring what we really want to measure?* To meet the "test of construct validity" the researcher has to focus on two important steps (adapted from Yin, 2002, p.35):

- Select and clearly define the specific concepts to be measured.

- Demonstrate that the "measuring instruments" that are going to be used, do measure those specific concepts.

The three tactics that appear in Table 10 to enhance the construct validity in a Case Study were used in this investigation; they are: multiple sources of evidence to analyse the same concept (data from documentation, archival records, interviews and direct observation), a chain of evidence was established, and key informants (experts in the area) reviewed the draft case reports (these tactics become clearer in the data analysis section).

Internal Validity

Issues of internal validity exist only when causal relationships are intended. Consequently only explanatory case studies have to deal with this problem. Most of this investigation is descriptive, therefore very few "suggestions as to causality" (also called associations or explanatory relationships between concepts or categories) are planned and no "definite causal" relationships (also called genuine causality or causal relationships between concepts or categories) are projected. However some inferences are made during the discussion. *To infer* is the act or process of deriving logical conclusions from premises known or assumed to be true. Some questions that can help to increase the quality of the inferences follow: "Is the inference correct? Have all the rival explanations and possibilities been considered? Is the evidence convergent? Does it appear to be airtight?" (Yin, 2002, p.36).

The other tactics that appear in Table 10 (pattern matching, explanation building, address rival explanations and use logic models) were taken into account and some aspects of them were used during the data analysis to increase the validity of the inferences.

External Validity or Generalization

Generalization in case studies is closely related to what was explained before about the different logics used when selecting cases in different kinds of research strategies. Replication logic or purposeful sampling is used in case studies as opposed to the sampling or "statistical" logic used in surveys.

When a survey is done the researcher does a statistical sampling because the intention is to do a statistical generalization. But when a case study is done the researcher does a purposeful sampling because the intention is to generalize to a theory (also called analytical generalization).

Analytical (or theoretical) generalization is the same logic utilized when generalizing the results of a specific experiment. When an experiment is carried out within a specific and controlled environment, its results are only applicable to that phenomenon in the same environment. This is not a problem because the original idea was not to generalize the results to the population but to generalize to theory. The same happens with the case studies. However in case studies and experiments, generalization is not automatic as it is in surveys. To further generalize to theory, more experiments or case studies (replications) must be done.

Consequently, the specific results of this research were generalized only to similar farms in similar conditions. Nevertheless, the general results and conclusions could be taken into account to build theory or support existing theory.

Reliability

An investigator passes the test of reliability when a later researcher can repeat the same investigation and obtains the same results as the earlier researcher. The objective of this test is to minimize errors and biases in a study by giving other people (a third party) the opportunity to check the results by repeating the research. Yin (2002) suggested two tactics to increase the reliability of a case study research (see Table 10):

- All the procedures should be documented by the utilization of data collection and data analysis protocols.

- The raw data of each farm should be organized in a case study database available for other researchers in case they wanted to repeat the analysis.

The organization of a database for each case study will also give other researchers the possibility to differentiate the data collected from the analysis made by the researcher. And it will also enable the present researcher to go back to the raw data in the future if necessary.

5.2 Data Collection

Most of the data was collected in a journey by the researcher to Argentina. The researcher first contacted the farmers and the experts by electronic mail or by telephone and invited them to participate in the research project. The conditions were explained and appointments were set with the farms and persons that accepted. Then the researcher travelled to Argentina and collected most of the data in a period of 20 days. After returning to New Zealand the researcher kept in contact with the farmers and experts in order to complete the information and clarify several points.

5.2.1 Main Data Collection Techniques

Six techniques were suggested by Yin (2002) to be the most commonly used in doing case studies. These techniques are similar to those proposed by Mintzberg (1973) for studying managerial work and the ones proposed by Rougoor, Trip, Huirne, & Renkema (1998) for studying managerial capacity:

- Documentation

- Archival records

- Interviews

- Direct observation

- Participant-observation

- Physical artefacts

In this research mainly four of these techniques were used to study the cases: documentation, archival records, interviews and direct observation. The study of the context was done through relevant literature and interviews with experts.

Documentation:

There are two different types of documentation, the primary source and the secondary source documentation (Mintzberg, 1973). The primary source documentation is that prepared by the dairy farmer himself or by people within the business. For example written plans, contracts, letters, memoranda, agendas, written reports of events, proposals, formal studies, etc (adapted from Yin, 2002 and Rougoor et al., 1998). People outside the dairy farm, in contrast, set up the secondary source documentation. Examples of this are tax data, accounting data, reports prepared by consultants in different areas, etc (Rougoor et al., 1998).

From this source of evidence very little information was collected. Only the accounting data of one of the farms was collected (*Farm 7*). The farmers from their archival records provided all the other physical and financial information.

Archival Records:

Relevant physical and financial information was collected from the last seasons on most farms:

Farm 1: Complete physical and financial information of the seasons July 2001-June 2002 and July 2002-June 2003 was provided. Also complete physical information for the period July 2003-June 2004.

Farm 2: Complete physical and financial information of the last three seasons (periods February 2001-January 2002, February 2002-January 2003 and February 2003-January 2004) was provided.

Farm 3: The Company has two dairy enterprises. Complete physical and financial information of the last three seasons (July 2001-June 2002, July 2002-June 2003, and July 2003-June 2004) was collected from one of the dairy enterprises. From the second dairy enterprise only the information of the last season was provided.

Farm 4: Complete physical and financial information of the last three seasons (periods July 2001-June 2002, July 2002-June 2003 and July 2003-June 2004) was provided.

Farm 5: Complete physical and financial information of the seasons February 2001-January 2002 and February 2002-January 2003 was provided. Also some physical and financial information of the period July 2003-June 2004 was obtained.

Farm 6: Only some general physical and financial information from the last seasons could be obtained.

Farm 7: Complete financial information was obtained from the last three seasons (periods July 2001-June 2002, July 2002-June 2003 and July 2003-June 2004). Some physical information was also obtained but not the all information needed.

Interviews:

The interviews were focus interviews also called semi-structured. This kind of interview follows a certain set of questions, is open-ended and assumes a conversational manner. The interviews were recorded, with permission from the interviewees.

Fourteen interviews of between two and three hours were done: one with each of the seven farmers or CEOs of the Argentine farms; two interviews with experts in the Argentine dairy sector and five more within the case study farms.

In the interviews with the Argentine dairy farm owners and CEOs and with other people working on the farms the researcher asked about the following topics:

- Their opinion and perception of 14 New Zealand innovations with potentiality to be adopted in Argentina.

- Then questions were asked about the adoption or rejection of the innovations in the farm and about the reasons for having adopted or rejected them.

- Additionally some questions were asked about the impact of the innovations on the dairy system.

- Finally some questions about the possibilities and constraints for adopting New Zealand innovations to Argentine dairy farms.

The two experts were asked about their opinions and perceptions of the same list of New Zealand innovations for Argentine dairy farms. They also were asked about their area of expertise and about their experience in the Argentine dairy sector. Additionally, they were enquired of about adoption of New Zealand innovations to Argentina in the past. Finally some questions were asked about the possibilities and constraints for adopting New Zealand innovations to Argentine dairy farms.

Direct Observation:

A field visit was made to each case study farm. The intention was to quickly inspect some specific paddocks, the main buildings, the plant and machinery, and the animals and also meet some of the people working on the farm. This observation was very important because it gave not only a general impression of the business but also specific and relevant information about whether the observer was correctly prepared or had some experience in the field.

Figure 53: Data Collection

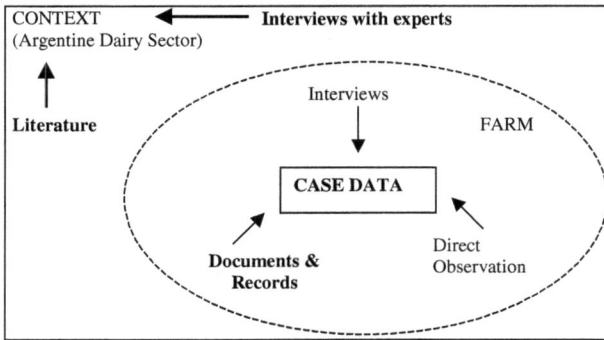

5.2.2 Principles for data collection

Three principles for data collection were taken into account (2002) as a way to maximize construct validity and reliability of the investigation. These three principles have been summarized by Rowley (2002) in the following way:

"**Triangulation (multiple sources of evidence)** – one of the great strengths of case studies as compared with other methods is that evidence can be collected from multiple sources. Triangulation uses evidence from different sources to corroborate the same fact or finding.

Create a case study database – a case study database of the evidence gathered needs to be collected. Whilst a report or dissertation may be the primary distillation of the case study, a further outcome which strengthens the repeatability of the research, and increases the transparency of the findings is a well organized collection of the evidence base. This base may include case notes made by the investigators, case study documents that are collected during a case study, interview notes or transcripts, and analysis of the evidence. When preparing a dissertation it will be useful to agree with a supervisor whether some elements of this evidence base should be presented as appendices to the dissertation.

Maintain a chain of evidence – the researcher needs to maintain a chain of evidence. The report should make clear the sections on the case study databases that it draws upon, by appropriate citation of documents and interviews. Within the database, it should be clear that the data collection followed the protocol, and the link between the protocol questions and the propositions should be transparent." (Rowley, 2002, p.23)

5.2.3 Perception Questionnaire

As was previously explained in the section about the Diffusion Theory, the way in which the potential adopters perceive an innovation can partially predict the rate at which that innovation will be adopted. To have an idea of the perception of the case study farmers, the experts and other farmers, a questionnaire was designed.

The questionnaire comprised 14 parts in which each of 14 New Zealand innovations were briefly explained. For each innovation the person was asked to mark from 1 (very low) to 4 (very high) for four characteristics. These were, "Level of Advantage", "Level of Compatibility", "Level of Trialabilty", and "Level of Awareness of Results".

In total 29 questionnaires were answered by three different groups of Argentine people: 1) 14 persons in the seven selected Argentine dairy farms, 2) five Argentine farmers that can be considered neutral about the New Zealand innovations, and 3) nine consulting officers of an Argentine institution of discussion groups. A summary of the results obtained is presented in the first chapter of results (Chapter 7).

5.3 Data Analysis

As Rowley (2002) stated "there are no cookbook procedures that have been agreed for the analysis of case study results, but good case study analysis adheres to the following principles:"

"The analysis makes use of all the relevant evidence.

The analysis considers all of the major rival interpretations, and explores each of them in turn.

The analysis should address the most significant aspect of the case study.

The analysis should draw on the researcher's prior expert knowledge in the area of the case study, but in an unbiased and objective manner" (Rowley, 2002, p.24, is a summary of Yin 2002, p.137-138).

5.3.1 The Context

Some characteristics of the Argentine socio-economic environment and the main characteristics of the Argentine dairy sector were described. To complement data from the literature two experts were interviewed and the previously explained perception questionnaire was completed by some Argentine farmers, consultants and experts.

The interviews were analysed by transcribing the relevant ideas from the tapes to a word processor and ordered by the topics of interest. Then the relevant ideas regarding the Argentine dairy sector were translated to English and reported into the first chapter of results (Chapter 7) of the present report.

5.3.2 The Cases

Qualitative and quantitative data were obtained from each farm.

The qualitative data collected through the interviews, the questionnaires and the visit to the farm was analysed by selecting the important information for the topic studied. The interviews were analysed in the same way as described before for the experts' interviews. The data from the interviews was complemented by the data of the questionnaires and the visit to the farm.

The quantitative data from the farms was analysed with the IFCN spreadsheet. The data from each farm from the 2002 and 2003 seasons was processed (excluding *Farm 6*). The main indicators that resulted from the spreadsheet analysis were then presented and analysed in the second results section of the present report (Chapter 8).

Follows an explanation of the main modifications done to the financial and physical data collected from the farms in order to make them comparable and in order to introduce it in the IFCN database:

Farm 1: Seasons 2002 and 2003 were loaded to the IFCN database. The financial data was originally in nominal AR-$ for each period and all the figures excluded GST. Only the data from the dairy enterprise was used; this enterprise bought the replacement heifers and the concentrates and silages from other enterprises of the same company. The farm provided the prices at which the concentrates and silages were bought. The cropping enterprise of the farm sold the concentrates and silages at market prices. The prices of replacement heifers were taken from the IFCN database and were the same for all the farms. *Farm 1* was the only farm in which the raising of the heifers was not analyzed as part of the dairy enterprise.

Farm 2: Seasons 2002 and 2003 were loaded to the IFCN database. The financial data was originally in nominal AR-$ for 2003 and all the figures excluded GST. However the financial data for the 2002 season was originally adjusted by the wholesale price index, therefore was converted again to nominal AR-$ in order to be compared to the other farms. Only the data from the dairy enterprise was used; this enterprise raised all its own replacement heifers and bought the concentrates from the cropping enterprises of the same company. The farm provided the prices at which the concentrates and silages were bought. The cropping enterprise of the farm sold the concentrates and silages at market prices.

Farm 3: The data had the same characteristics to *Farm 2's* data; therefore the same modifications were done.

Farm 4: Seasons 2002 and 2003 were loaded to the IFCN database. The financial data was originally adjusted by the wholesale price index, therefore was converted again to nominal AR-$ in order to be compared to the other farms. The original data had the same characteristics to *Farm 2's* data in respect to replacement heifers and concentrates bought.

Farm 5: Only the 2002 season was analysed. The financial data was originally in nominal AR-$ for 2003 and all the figures excluded GST. The original data had the same characteristics to *Farm 2's* data in respect to replacement heifers and concentrates bought.

Farm 6: Only some general physical and financial information from the last seasons could be obtained.

Farm 7: The financial accounts from 2002 and 2003 were provided. They were in nominal AR-$. The original data had the same characteristics to *Farm 2's* data in respect to replacement heifers and concentrates bought. This was the only farm with unpaid family labour, the wages per hour of family labour provided by the IFCN were used to calculate the total labour costs.

6   NEW ZEALAND INNOVATIONS FOR ARGENTINA

6.1   Selection of the New Zealand innovations

The selection of the New Zealand innovations for the present study was based on three sources: the knowledge and experience of Argentine farmers who are adopting New Zealand practices and technologies; relevant literature and articles in the rural press that describe the typical practices and technologies of both countries; and the knowledge and experience of the researcher on Argentine and New Zealand dairy farm systems.

The criteria for the selecting the New Zealand innovations was as follows:

- The principles, practices and technologies have to be typical[18] of New Zealand dairy systems and not common on Argentine farms.

- The innovations must have been adopted, at least partially, on one Argentine dairy farm.

- The innovations selected have to cover the main aspects of a dairy system.

Some of the principles, practices and technologies that were chosen as New Zealand innovations may be also found in other countries, or may have been suggested as recommended dairy farming practices by researchers and developers in other parts of the world. However in New Zealand, these principles, practices and technologies are actually being used and considered normal and typical.

Initially, the researcher identified a list of 14 practices and ideas that matched the above criteria. This list was sent to two of the most renowned Argentine dairy farmers who have adopted New Zealand innovations. Both of them approved the list and suggested some small changes. In New Zealand, Nicola Shadbolt and Colin Holmes, who are

---

[18] Note that 'typical' means characteristic of New Zealand farms, which is not the same as more frequent. For example, 'formal pastoral budgets' are not used in most New Zealand farms (Gray, 2001) but they are recommended by most New Zealand consultants and used by many top New Zealand farmers (Dexcel, 2003) and are probably more spread throughout New Zealand than any other country in the world (Hodgson, 1990; Holmes et al., 2002).

experts on New Zealand dairy farm business and systems respectively, concurred that the innovations suggested were typical for New Zealand farmers.

In addition, in order to include all the relevant New Zealand innovations, during the data collection, the farmers and experts interviewed (14 people within the seven case studies farms and two experts in the Argentine dairy sector) were asked about the group of 14 innovations selected. Most of them agreed that the list was complete and that it covered the main points of a dairy system.

Once the data had been collected, it was decided that four of the innovations were not sufficiently relevant or that they could be combined with other innovations, therefore the analysis includes 10 New Zealand innovations.

Table 11: Selection of New Zealand Innovations for Argentina

Original Innovations Proposed (14)	Final Selection of Innovations (10)
1) Stocking rate decided taking into account prices of milk and supplements	1) Focus on production per hectare
2a) Intention to raise the phosphate level of soils of the farm	2) Marked importance to pasture production
2b) Intention to have pastures that last more than 4 years without the need to replant them	
2c) Utilization of adequate species on pastures	
3a) Quantitative pasture monitoring	3) Quantitative pasture monitoring
3b) Target pre and post grazing pasture covers	
4) Utilization of formal pastoral budgets	4) Utilization of formal pastoral budgets
5) More ´technology´ and less working hours	5) Less than 15 cows per set of teat-cups, and other innovations that impact on labour productivity
6a) Skilled and motivated people working on farms	6) Skilled and motivated people working on farms
6b) Working contract similar to New Zealand share-milking contracts	
7) Seasonal calving, one or two calving periods	7) Seasonal calving, one or two calving periods
8) New Zealand genetics	8) New Zealand genetics
9) Rearing of calves in groups	9) Rearing of calves in groups
10) Style of milking shed and milking system	10) Style of milking shed and milking system

6.2 Definition of New Zealand innovations for Argentina

In this section each innovation is defined by differentiating the typical Argentine practices from the New Zealand innovations. Additionally, the indicators and criterions that were used to assess the acceptance or rejection of the New Zealand innovations on the case studies are defined.

All the innovations selected are integral parts of a system. They were separated here in order to better understand and differentiate one from the other. The sequence in which they are presented below runs from general to particular and from principles to technologies.

1) Focus on Production per Hectare

Milksolids produced per hectare is the ultimate measure of physical performance in New Zealand dairy farms (Holmes et al., 2002). New Zealand dairy farmers and researchers have been traditionally focused on maximizing milk production per unit of area; the management of stocking rate has been one of the key drivers of productivity of the farm and land (MacDonald et al., 2001; Penno, 1999).

In New Zealand there is an ongoing debate between farmers who prefer to have higher production per cow and a lower stocking rate, and other farmers that prefer exactly the opposite. Even though farmers who have a preference for lower stocking rate, and are targeting higher per cow production, may seem to be focused on cow productivity, they are ultimately interested in producing as much milksolids per hectare as possible, the only difference is that they prefer to do it with a lesser number of cows.

In Argentina, as in New Zealand, land is more than half of the total assets of the typical dairy farm. It comprises more than 55% of the total assets in New Zealand typical farms and more than 75% in Argentine typical farms (IFCN, 2003). Therefore land can be considered as the main asset of the dairy farm business. Despite this fact the focus on production per hectare is not so clearly defined in Argentina where researchers and typical farmers seem to be more focused on productivity than in New Zealand (Castignani & Zehnder, 2003). Very often in Argentina the cow productivity is seen as an end in itself; several farmers have a target milk production per cow but do not know how much they should be producing per hectare (researchers experience in Argentine dairy farming).

Accordingly, a clear *Focus on production per hectare* is considered an innovative principle for typical Argentine dairy farms. To assess the adoption or rejection of this innovation the case study farmers were asked questions in order to assess if they were more focused on productivity per hectare or productivity per cow.

To follow this principle, without necessarily increasing the level of supplements, New Zealand systems try to maximise the amount and quality of feed produced per hectare and optimise the utilisation of that feed by the cows (Holmes et al., 2002).

2) Give Marked Importance to Pasture Production

In New Zealand there is a clear focus of farmers on trying to maximize the productivity and vigour of pastures in their farms. Farmers, oriented by research, use specific levels of fertiliser inputs, and utilise certain grazing techniques in order to maximise growth, quality and amount eaten from their pastures. Phosphate fertiliser is spread annually in order to keep recommended levels of phosphate in the soils (Thomson, Roberts, McCallum, Judd, & Johnson, 1993). Additionally, farmers look after their pastures, trying not to graze them too intensively in order to promote re-growth (especially in winter). They also do "on-off grazing", "24 hour grazing", "square breaks", and other practices to prevent damage to pastures by the cows on wet days (Clark, Carter, Walsh, Clarkson, & Waugh, 1994; Pande, Valentine, Betteridge, MacKay, & Horne, 2000; Thomson & al, 1993). Very often New Zealand farmers are recommended to give priority to their pastures over their cows in situations in which they have to decide which of the two would lose condition the most (Lambourne & Betterridge, 2004). Normally in New Zealand the durability of pastures is not an issue and they last several decades without the need for re-sowing.

In contrast, pastures very rarely last more than four years in Argentina. The causes are unknown, but it is not common to fertilize pastures or to leave optimal residuals for re-growth, which must contribute to this lack of longevity. In Argentina a typical farmer would give priority to the cows over the pastures in situations in which one of the two have to lose condition in some degree (interviews with experts, farmers and consultants).

Therefore the principle of *Focusing on pasture production* was considered to be the second New Zealand innovation for Argentine dairy farmers. The indicators used to assess the adoption of this innovation by the case study farms were mainly four:

- Ask about the actual levels of phosphate in soils and if they had the intention to raise the phosphate levels to those recommended by experts to be more suitable for optimum pasture production.

- Ask them if they were doing any kind of special management to look after the pastures in order to increase the durability of the pastures (for example leaving minimum residuals at certain times of the year, or preventing cows from damaging the pastures during wet conditions).

- Ask if they were actively looking for pasture species that better suit their production systems and ecological conditions.

- Assess whether they would give priority to the cows or the pastures in the case that they have to choose which would have to lose condition.

As previously stated, to maximise milk production per hectare (first New Zealand innovation) it is important not only to produce as much grass as possible (second New Zealand innovation) but also to utilise efficiently the grass produced. Two New Zealand practices that assist farmers to better utilize their pastures were defined as New Zealand innovations to Argentine farms: *Quantitative Pasture Monitoring* (third New Zealand innovation) and *Utilisation of Formal Pastoral Budgets* (fourth New Zealand innovation).

3)   Quantitative Pasture Monitoring

Pasture is the main feed of dairy cows in typical New Zealand and Argentine dairy farms (IFCN, 2002). Therefore, efficient grazing management is relevant for farms in both countries.

It is well known that New Zealand farmers rely more on pasture than probably any other farmers in the world (Mitchell, 2002). This situation is generated by, among other things, the fact that pasture is by far the cheapest feed in New Zealand (Holmes, 2003a).

Over the years, with the help of research and extension, New Zealand farmers have developed effective pasture management skills (McCall & Sheath, 1993).

For reasons that are beyond the scope of the present study, Argentine farmers, researchers and consultants have not been so interested in pasture management as their counterparts in New Zealand (researcher's experience in Argentina and interviews with several Argentine dairy farmers and experts). However, to improve the general efficiency of Argentine dairy farms, all the resources must be well managed and grass is an important resource for Argentine farmers. In this context, Argentine farmers may benefit from research and best practices of farmers in New Zealand.

Professor John Hodgson, a renowned scientist in grazing management both in the United Kingdom and in New Zealand, stated that "the regular appraisal of sward conditions is an important prerequisite of efficient grazing management" (Hodgson, 1990, p.181). He also suggested that the adoption of effective recording programs could greatly increase management objectivity.

It is common to find farmers in New Zealand who regularly measure the pasture cover (in kilograms of dry matter) of their paddocks and farms, and record the measurements on paper or computer spreadsheets (Hodgson, 1990). This quantitative recording year after year gives them an estimate of the total pasture production of a typical season. They can use the information to approximately assess at which times of the year the pastures growth rates are higher or lower. The ability to assess the cover of their pastures at a certain moment in time enables them to set pasture cover targets for different times of the year. Additionally, they can decide pre and post grazing covers based on research that investigated the optimal covers of pastures at different times of the year.

Finally, the recording of pasture cover through several seasons allows farmers to do feed or pasture budgets. The accuracy of the estimations done on the budgets depends on the quality of the data recorded in the previous years.

To assess if the Argentine farmers were quantitatively monitoring and recording pasture they were asked if they actually have a recording programme for pasture growth (or pasture cover) during the whole season.

4) Utilisation of Formal Pasture Budgets

The last New Zealand innovation related to the principle of maximizing production per hectare is the *Utilisation of formal pasture budgets.*

As with any kind of budget (for example financial budgets), pasture budgets are a balance between demand and supply, outputs and inputs. The demand for feed is given by the total requirements of the animals on a farm for a certain period of time in the future. And the supply is given by the estimated average growth of the pastures of the farm during the same period of time. Pasture budgets are useful to anticipate periods of pasture surplus and deficit, and to allow farmers to plan in advance (Hodgson, 1990; Hopkins, 2000).

Some farmers do pasture budgets mentally; they do not use paper or computers to calculate the differences between pasture growth and animal requirements (Gray, 2001). In the present study, only feed budgets that are recorded on paper or computer (formal pasture budgets) were considered to be New Zealand innovations because this provided a basis for assessment of the extent to which the innovation has been adopted.

5) Skilled and Motivated People Working on Farms

In the present study "people working on farms" means people that live on and are in charge of the daily activities of the dairy farm. They are people that work on the farm and do the activities that are inherent to a dairy farm (feeding, milking, and looking after the health and reproduction of cows).

One of the strengths of the New Zealand dairy sector is that most people on farms are energetic, positive, technically skilful and optimistic (Holmes, 2003b). This is a very important characteristic of the New Zealand dairy sector and probably one of the drivers of its success.

In New Zealand, most dairy farms (63% of all herds) are owner-operated; owner operators are farmers who either own and operate their own farms, or who employ a manager to operate the farm for a fixed wage. Managers employed by landowners were less than 1% of all owner-operated herds in 2001 (LIC, 2001/02), therefore more than 60% of all New Zealand dairy farms are operated by their owners. Additionally, people working on farms are at same time the owners of the main dairy company (see section 3.1.4).

In Argentina, by contrast, the levels of skills possessed by staff working on farms have been suggested as one of the main weaknesses of the Argentine dairy sector (Guardini, Labriola, & Schaller, 1999). This fact has an impact on the performance of the Argentine dairy systems and also on the Argentine dairy sector in general.

Accordingly *Skilled and Motivated People Working on Farms* was defined as a New Zealand innovation for Argentine dairy farms. However the level of skilfulness and motivation of the people working in the case studies was very difficult to assess in such a limited time (20 days of data collection). Motivation is very related to ownership and ambition; therefore the owners and the employees that want to grow in the dairy business are usually the most motivated people. In addition, people can be more motivated to work towards increasing a dairy farm's performance if they are associated to that performance. Skilfulness, on the other hand, is more related to education and training.

Several indicators were used to assess the adoption of this New Zealand innovation:

- A meeting with the main operator of one dairy farm (for each case study farm) in order to assess his motivation and skilfulness.

- The level of education of people working on farms (for example, years of formal education before start working on the job, specific formal training for the job, and other).

- Existence of "on the job training" by people of the farm that have more experience or knowledge than them.

- Observation of the working environment (technology provided, milking shed characteristics, roads and races, and other details).

- The job conditions (working hours per day and per year, milking times, time off and holidays, and wages) were related in some degree with the level of skilfulness. It was assumed that farmers who are looking for more skilful people offer better job conditions.

- Assessment of the possibilities given to the people working on the farm to grow in the dairy farm business (for example, by increasing the level of responsibility when they are more experienced, by participating in the results of the farm).

6) Less than 15 cows per Set of Teat-cups, and Other Innovations that Impact on Labour Productivity

In New Zealand dairy farms there is approximately one person for each 100 to 140 cows (IFCN, 2002). This relationship of cows per person is the highest in the world (IFCN, 2002). In order to complete all the daily activities, New Zealand farmers have to be very efficient with their time. They accomplish this by utilizing adequate technologies and also by designing relatively simple systems.

Every day, during the time when the cows are lactating, the dairy farmer has to milk the cows. This job is done for approximately four hours per day for 250 to 300 days a year (Holmes et al., 2002). "Milking, is therefore a major factor in the work routine on all dairy farms" (Holmes et al., 2002, p.155). The number teats-cups has a big impact on the total milking time, the more sets of teat cups the faster it is to milk the same number of cows; up to a limit (usually 25 set of teat-cups per person) in which the person can not attend so many sets of teat-cups. Therefore the number of cows per set of teat-cups has a big impact in the labour productivity of the farm (see Holmes et al., 2002). In New Zealand typical farms 15 cows, or less, are milked per set of teats-cups. This relationship is relatively low in comparison to most of dairy farm systems around the world (IFCN, 2002).

Other technologies normally utilised are, for example, motorbikes, front loaders on tractors, posthole borers, automatic teat spray, automatic effluent spreaders, motorised backing-gates, overhead backing gate, and automatic systems to wash the collection yard. Additionally many dairy farms in New Zealand use contractors to do the main "non-dairy" work (for example, silage conservation, cultivation, fencing and farming development), this increases the labour productivity of the dairy farmers.

In Argentina, on a typical farm, there is one person for each 40 to 60 cows (IFCN, 2002). Dairy farmers in Argentina spend most of their working time milking the cows because normally more than 25 cows are milked per set of teat-cups. Instead of the other technologies used in New Zealand, horses are used instead of motorbikes and a

115

characteristic example is that usually people use spades to dig postholes for fences instead of using posthole borers as in New Zealand (researchers experience in Argentine dairy farms). In Argentina, for all previously mentioned things, typical farms are more labour intensive than those of New Zealand.

To assess the adoption or rejection of this innovation, four aspects were assessed:

-   Number of milking cows per set of teat-cups.

-   Utilization of other technologies as motorbikes, front loaders on tractors, posthole borers, automatic teat spray, automatic effluent spreaders, motorised backing-gates, overhead backing gate, and automatic systems to wash the collection yard.

-   Simplicity of the systems, as for example simplified milking routine.

-   In addition the farm owners were questioned about their preferences for either a system that employs more labour and usually invests less in plant, machinery and buildings; or a system that employs less people but invests more in plant, machinery and buildings.

7)   Seasonal Calving, One or Two Calving Periods per Year

On the majority of New Zealand dairy farms, the whole herd calves during a period of about 12 weeks in late winter and early spring (Garcia & Holmes, 1999; LIC, 2002/03). This is done in order to match the daily feed requirements of the herd with the daily pasture growth, which is the main feed for most of New Zealand herds. Consequently a big proportion of the pasture grown is harvested directly by the cows at the lowest possible cost (Garcia & Holmes, 1999; Holmes et al., 2002; Mitchell, 2002).

In marked contrast, nearly all the dairy farms in Argentina have some cows calving in every month (Garcia, 1997). The main cause of this is probably the fact that most of the milk produced is used to supply the local Argentine market which needs a relatively constant supply of milk during the whole year (Gutman et al., 2003). Consequently, most of the Argentine dairy companies encourage winter and summer milk production, when feeding is more expensive, by raising the price of milk (interviews with farmers, experts and consultants). But despite this constraint some Argentine farmers have adopted seasonal calving systems successfully, and some of them still produce most of their milk during the periods in which the milk price is higher (examples are all case study farmers interviewed apart from *Farmer 4*).

To assess whether the Argentine farms were calving seasonally or all year round, the farmers were asked about their herds' calving period. Calving periods shorter than four months were considered seasonal. Those farms with split calving were considered seasonal only if both calving periods were also shorter than four months each.

8)   New Zealand Genetics

New Zealand cows are known for having been selected under conditions of seasonal pastoral dairy systems. Therefore these cows are able to conceive every year, and are prepared to walk long distances and harvest grass as their main feed (LIC, 2001). The fact that large numbers of cows are milked per person, in comparison to other countries (IFCN, 2002), ensures that cows that need special treatments (for example, need udder stimulation before milking, have bad temper or are slower to get milked) have traditionally been culled (Holmes et al., 2002). More recently, longevity, udder and body conformation, and reproductive performance have been introduced to the selection criteria (LIC, 2001). Another characteristic is that New Zealand cows produce more concentrated milk, which is preferred by the New Zealand dairy companies that are oriented mainly to overseas markets (LIC, 2001).

In Argentina dairy cows are mainly Holstein, but some cows have a high degree of North American Holstein while at the other extreme some cows have a degree of old Friesian genetics (ACHA, 2000). There is a trend to increase the genetic merit of Argentine Holstein cows, mainly by utilisation of North American bulls (researcher's experience in Argentine dairy farming and interviews with farmers). North American cows have been selected to produce high volumes of milk, mainly in confinement systems where they are fed with total mixed ration (TMR). They can consume very high amounts of these diets and convert them efficiently to large volumes of milk (Mayne, 1998).

The criteria used to assess if the Argentine farms were using New Zealand genetics was simply by asking farmers if they were using New Zealand genetics, and also what proportion of the cows and heifers were being inseminated with New Zealand semen.

9)   Rearing of Calves in Groups

On most dairy farms in Argentina calves are reared in individual cages or tied to a chain and fed with milk or milk substitute in buckets. Each calf has two buckets, one for the milk or milk substitute and the other for water. The person that feeds the calves fills the buckets once or twice a day. Occasionally the calves are moved to keep the ground or pasture under the calf clean (AACREA, 1986).

Conversely, calves are reared in groups in small paddocks on New Zealand dairy farms. Instead of drinking the milk or substitute from buckets, groups of calves suck through "calf-teats" from a container; a barrel of 200 litres can be used (Holmes et al., 2002).

To assess if the Argentine farmers were doing *Rearing of Calves in Groups* they were asked which calf rearing system they were using for most of their calves.

10)   Style of Milking Shed and Milking System

The most typical New Zealand milking sheds (also called "farm dairies" or "cow sheds") have one concrete wall and are open on three of the four sides. The most common milking systems are the high-line single herringbone parlours (also called "swing over" parlours). This system usually has a central high pipe with sets of teat-cups hanging from it (Holmes et al., 2002). These sheds and systems are designed so that the cows can be brought from paddocks and milked in the least possible amount of time, and then returned to the paddocks again. They are characterized by their simplicity and relatively low cost. Nowadays rotary milking sheds are becoming increasingly popular, especially on bigger farms (more than about 500 cows).

In North America, milking sheds are usually more complex, most of them are built in similar way to houses and offer more protection against the weather conditions than in New Zealand. Usually they have two pipe lines with sets of teat-cups (IFCN, 2002). In North America cows usually spend more time inside the milking shed than in New Zealand because they produce more milk per cow and because the milking routine includes more tasks, and is longer. There are also rotary milking sheds on some North American dairy farms (IFCN, 2002).

Argentine milking sheds are somewhere intermediate between those of New Zealand and North America. Typical Argentine milking sheds have two walls, two pipe lines and sets of teat-cups (IFCN, 2002). In general cows spend more time in the milking shed in Argentina than in New Zealand because the milking routine is more complex but after the milking they return to pastures as in New Zealand (AACREA, 1986; Holmes et al., 2002).

Direct observation of the milking shed and surroundings was used to classify the Argentine farmers into those that utilise or not the New Zealand *Style of Milking Shed and Milking System.*

7 RESULTS (Part 1)

The Results of the present study are presented in two chapters. In the first chapter (Chapter 7) the case studies are presented and issues related to the adoption or rejection of New Zealand innovations are addressed. In the second chapter (Chapter 8) the performance of the case studies in 2002 and 2003 is analysed and the impact of the innovations adopted is assessed.

7.1 The Case Study Farms: Adoption and Rejection of New Zealand Innovations

The seven farm businesses chosen as cases studies are medium to large dairy farms by Argentine standards. They are larger in cow numbers and hectares than the Argentine average dairy farm (Gambuzzi et al., 2003); and larger than the typical farm chosen by the IFCN (2002) as the major contributor to Argentine milk production. Additionally, four of the case studies can be considered of a very large scale with more than 1,500 cows.

The cases are described from the largest to the smallest in number of cows for the last season analysed (2003).

For the complete description of each New Zealand innovation please refer to section 6.2.

In the methodology chapter in section 5.1.4 the way in which the case studies were selected is explained.

Table 12: Case Study Farms Outlook (season 2003)

FARMS	Total cows (average for the season)	Area utilised for cows and heifers (has)	Stock-ing Rate (cow per ha)	Cows live weight (kg)	Milk produced per year (1000 of kg ECM[19])	Milk produced per cow (kg ECM)	Milk produced per ha (kg ECM)	% of feed brought from outside the area	Approx market value of land (US-$)[20]	Total people working on the dairy enterprise (labour units)	Cows per person	Number of milking sheds	Cows per set of teat-cups
Farm 1	6,350	2,603[21]	2.44[3]	605	29,744	4,684	11,427[3]	43%	2,500	70.4	90	18	25
Farm 2	2,530	1,435	1.76	424	12,372	4,890	8,622	49%	1,400	25.7	98	3	25
Farm 3	1,700	1,665	1.02	500	8,839	5,200	5,309	39%	2,000	20.0	67	3	26
Farm 4	1,375	780	1.76	400	6,697	4,871	8,587	55%	2,000	25.5	69	1	24
Farm 5	915	391	2.34	442	3,750	4,099	9,593	63%	1,400	9.2	100	1	30
Farm 6	400[4]	250[22]	1.60[4]	520	1,954[4]	4,886[4]	7,818[4]	0%[4]	2,000	8	50	1	20
Farm 7	400	600	0.67	432	1,557	3,735	2,596	60%	1,400	3.4	119	1	25

121

[19] Energy Converted Milk (ECM) with 4% fat and 3.3% protein. Formula used for adjustment: $ECM = (milk\ production\ in\ litres * ((0.383 * \%\ fat + 0.242 * \%\ protein + 0.7832)/3.1138)$.
[20] Calculated based on the market value of land for the typical Argentine farms (IFCN, 2004).
[21] Farm 1 is the only farm in which the area for raising the heifers is not included, for all the other farms the hectares include all the area used for lactating and dry cows, plus the area utilised for rearing the heifers.
[22] The data for Farm 6 was estimated from data given during the interview with the farmer (no farm records provided).

7.1.1 Farm 1

General Characteristics

The Farm Business

Farm 1 is a dairy enterprise owned by a large company. This company is also the owner of more farms (dairy, cropping and beef farms) and some other agribusinesses. The company has a long tradition in dairy farming (since 1930) and is well known in the region where the dairy farm is situated. As was mentioned in the methodology chapter (section 3.1.4) no further details are given for confidentiality reasons.

Farmer 1 (see below) mentioned that *'...today what we want is to produce milk as efficiently as possible, at the lowest cost, with a pastoral system helped by the supplements that we have...'*

The Farmer

The person who was interviewed is the Chief Executive Officer (CEO) of the farming division of the company. For confidentiality reasons in this report he is called *Farmer 1*. This person is responsible (among other things that are not related to the present study) for the main tactical decisions of the dairy enterprise. He is also responsible for making most of the strategic decisions; however the approval of the owners of the company is needed for strategic decisions.

Farmer 1 is a cosmopolitan person who reads many farming-related journals from several areas of Argentina and other parts of the world. He has travelled to New Zealand, Australia and the United States of America to learn about other farm systems. Some aspects of the New Zealand dairy systems (some of them also found in Australia) have attracted his interest and have been implemented in some of the company's dairy farms.

Feeding Sources

- 57% of total annual requirements for the cows (lactating and dry) are produced on the milking platform (grazed lucerne and lucerne silage).

Figure 54: Feeds Eaten by *Farm 1* Cows

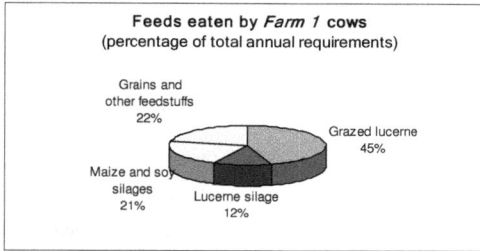

- The company produces most of the feeds that are used in the dairy enterprise; the cropping enterprise sells the grains to the dairy enterprise at market prices.

- "Other feedstuffs" are: maize grain, wheat bran, toasted soybean of broken or small beans, brewers' grains, cotton seeds, and sunflower expeller.

Adoption and/or Rejection of New Zealand Innovations

1)   Focus on Production per Hectare:

- *Farmer 1* mentioned during the interview that the company wants to maximize the production of milksolids per hectare.

- *Farmer 1* showed the researcher a report that he prepared for a seminar about production systems in which he describes the main areas that, according to his experience, affect the production per cow. These were: feeding, reproduction, health, and genetics. Then he related production per cow to stocking rate, as the two main factors that affect the production per hectare. The ultimate physical indicator for him was production per hectare. Therefore the focus of *Farm 1* is very similar of that of typical New Zealand farms.

- *Farmer 1* mentioned in the interview that the company tries to add value to its produce as much as possible. The dairy enterprise utilizes some of the soybean and maize grain from the cropping enterprise with the objective of converting it to milk.

- This innovation was completely adopted in *Farm 1* (mark=100%).

2)   Give Marked Importance to Pasture Production:

-   All pastures in *Farm 1* are of composed of lucerne (also called alfalfa) and white clover.

-   *Farmer 1* mentioned that they would like to increase the durability of their pastures.

-   The management of the pastures and the levels of fertilization used were typical for the region. No special management or fertilization was mentioned.

-   *Farm 1* is not really focused on pasture production. This innovation is just starting to be taken into account; at the moment of the visit to the farm this innovation did not seem to be an aspect in which they though they should improve too much (mark=25%).

3)   Quantitative Pasture Monitoring:

-   Each paddock with lucerne is grazed or cut between five and six times per year. It was a common practice in *Farm 1* to measure mass production of selected paddocks at certain times of the year. *Farmer 1* considered that this was not enough and that more measurements should be made.

-   *'This is probably the most lacking and problematic aspect for us'*, *Farmer 1* stated about pasture management. *'We should start by implementing the methodology of measuring grass'*. *'We are thinking of establishing a system in which university students could come to get farming experience and at the same time they would provide the service of measuring pasture cover for the firm'*.

-   This innovation is just starting to be taken into account for adoption in *Farm 1* (mark=25%).

4)   Utilization of Formal Pasture Budgets:

-   A yearly feed budget is drawn up; this budget is more focused on supplements and concentrates than on pasture.

- Cows usually eat all the available lucerne and the farmer tries to complete their requirements with supplements. In the periods in which the lucerne grows more than the cows' requirements, paddocks are closed, and it is conserved in the form of hay (bales) or silage (pit/stack).

- Formal pasture budgets are not used, they use a yearly feed budget; therefore, this innovation was not adopted in *Farm 1* (mark=0%).

5)   Skilled and Motivated People Working on Farms:

- During the visit to this farm it was not possible for the researcher to meet the main operator of one dairy farm or to see the working conditions. However, the researcher had previously visited the *Farm 2* years before and had the opportunity to visit some of its dairy farms and to talk with some of the farmers. The CEO confirmed that there were few changes in the labour structure since that visit.

- *Farm 1* has quite a vertical labour structure. There is a "management team" that is formed by the CEO and a group of veterinarians. Each veterinarian is in charge of a group of dairy farms (approximately 6 farms each) and is responsible for making the main tactical decisions in those dairy farms. In charge of the basic activities of each farm there is a dairy farmer (for whom the main activity is the milking of the cows) and his family. Additionally, in some of the larger dairy farms within *Farm 1*, some workers are employed.

- In terms of formal education, the CEO and the veterinarians have at least 5 years of tertiary education. Most of the dairy farmers finished primary school (first 7 years of formal education) and some of them also finished secondary school (5 more years); but usually they had not had any formal education in agriculture and they have not been specially trained on *Farm 1* during their periods of employment.

- The job conditions on *Farm 1* seemed to be quite similar to most Argentine dairy farms, which are as follows: a normal working day has between 8 and 10 hours; it is common to have two days off every two weeks and sometimes two days off every four weeks. Holidays range between 10 to 20 days per year. The cows are milked at 3 a.m. and 3 p.m.; each milking takes approximately four hours. The dairy farmer and his family usually get approximately 12% of the milk returns, and he usually pays for extra labour if it is needed.

- In *Farm 1*, there are five walkthrough milking sheds in a total of 18 dairy farms. This type of old milking shed can still be found in some of the most traditional dairy farm areas of the country. They are uncomfortable for the people that milk the cows and consequently they are not longer built in Argentina.

- *Farm 1* has a vertical labour hierarchy; the people that are performing the daily tasks do not have formal farming training or education, and the job conditions are relatively hard. In conclusion *Farm 1* had not adopted this New Zealand innovation and it is relatively typical in this respect (mark=0%).

6)   Less than 15 cows per Set of Teat-cups, and Other Innovations that Impact on Labour Productivity:

- In *Farm 1*, there are between 22 and 25 cows milked per set of teat-cups. In New Zealand, typical farms milk 15 cows or less per set of teats-cups (see section 3.5.7).

- From what was defined as "other technologies" in section 3.5.7 only front loaders on tractors are used on *Farm 1*.

- The general impression of the interviewer was that in *Farm 1* they are not so interested in getting similar labour productivities to those of New Zealand. The management team takes the labour productivity into account and they try to increase it constantly but not by investing more in plant and machinery as in New Zealand. Instead one strategy is to simplify the system; for example, this farm utilizes the simplified milking routine that is typical in New Zealand farms. Another strategy is the specialization of tasks, which means that each person specialises in a limited number of tasks; for example, most dairy farmers in *Farm 1* spend most of their working time milking cows.

- In *Farm 1*, they had more than 15 cows per set of teat-cups. Because they had an average of 23.5 cows set of teat-cups, they use front loaders on tractors and a system that is relatively simple, it is defined that *Farm 1* adopted 45%[23] of this innovation.

7)   Seasonal Calving, One or Two Calving Periods per Year:

-   From the 18 dairy farms of *Farm 1*, 3 of them have seasonal calving with one calving period. The cows calve for approximately 12 weeks during February, March and April. They have chosen this calving period in order to produce milk during the period of higher milk prices (June-July, see section 2.6.2.3).

-   In *Farm 1*, the management team has carried out some research about the advantages and disadvantages that they find in adopting seasonal calving:

-   **The advantages that they found are:** events occurs in sequences and consequently only one task must be done at a time (calving, rearing of calves, mating and drying off); most of the cows have common requirements at a given time; increases the possibility of feeding the cows better in the periods that they have higher requirements (peak of lactation and mating); the possibility of producing a higher proportion of milk during the period of the year when the milk price is higher; possibility of choosing not to have any cow calving during the summer; better results are achieved on the raising of replacement heifers; better reproductive performance; possibility of choosing not to have cows producing milk during the warmest months of the year; possibility of producing more silage and hay reserves; the fact that there is a period of time in which the cows do not produce milk; facilitation of repair and maintainenance of the farm structure and organization of holidays for the staff.

-   **The disadvantages that they found are:** it is impossible to do seasonal calving if only one dairy farm is owned in Argentina because the dairy company requires milk on every day of the year; more labour needed at the beginning of the season; the need for bigger milking sheds to milk the same number of cows per year; and the need for more space to rear the calves.

---

[23] It was assumed that the higher the number of cows per set of teat cups the lower the adoption of this innovation. Less than 15 cows per set of teat cups was defined as nearly 100% adoption, more than 30 cows per set of teat cups was defined as nearly 0%.

- *Farm 1* adopted this New Zealand innovation partially, in three of its 18 dairy farms (mark=17%).

8) New Zealand Genetics:

- In *Farm 1*, they have been using New Zealand Jersey genetics in all heifers during recent years; actually approximately 20% of all cows and heifers are crossbreds. They decided to use Jerseys because they used to have a lot of calving problems in heifers and because they wanted to increase milksolids per litre of milk.

- Presently, they are thinking of continuing with the same practice (to cross all heifers with Jersey) but they are considering using North American Jersey instead. They believe that the North American Jerseys would have the same advantages of the New Zealand Jerseys and additionally increase the production per cow.

- They utilized New Zealand Holstein Friesian genetics on their Argentine Holstein cows and the decided not to use it anymore for several reasons: the resultant heifers had *'udder problems'* and also calving problems. In addition, the level of milksolids per litre did not change significantly in respect of the Argentine Holstein, and also they did not like the phenotype of the New Zealand Holstein Friesian.

- They have been trying New Zealand Jersey for some years now and they have found some advantages in using it:

- In a recent seminar given by the CEO of the dairy enterprise he showed how, for them, the Jersey cows produce more income per hectare than the crossbreds or Holstein Friesian cows. He showed that at the same stocking rate in kilograms of live weight per hectare (adjusted by metabolic weight), Jersey cows produce more milksolids per hectare. This is an advantage for them because the dairy company, to which they sell their milk, pays for milksolids (milk-protein + milk-milk-fat).

- One of the reasons why in *Farm 1* they started to use New Zealand Jersey genetics was because they were having calving problems. Using New Zealand

Jersey semen to mate their American Holstein heifers decreased the mortality of calves and the need to help cows to calve.

- The crossbreed cows, which resulted from using New Zealand Jersey semen on the North American Holstein heifers, proved to stay longer in the dairy farm because they had less fertility problems, less abortions, less lameness, and less mastitis. Further, the crossbreds showed better temperament, they were lighter and consequently caused less damage to pastures and they also recovered faster after calving than the North American Holsteins.

- The dairy company prefers milk with a higher proportion of milksolids because it costs less per kilogram of milksolids to be transported and is easier to process.

- *Farm 1* adopted this New Zealand innovation partially (mark=20%).

9) Rearing of Calves in Groups:

- In *Farm 1*, all calves are reared in the typical Argentine way.

- Some years ago in *Farm 1* they tried to rear some calves in groups and they had a very high level of mortality due to an epidemic of salmonella. They think that the problem could have been caused by continued use of the same area of land for rearing calves for too many years (since 1955). They designed a calf-feeder with a tank of 400 litres and approximately 30 calf-teats that could be fitted onto the front loader of a tractor. They think that the innovation could be useful for them because it take less time to feed the calves. They are planning to try this New Zealand innovation with some male calves again in the near future.

- This New Zealand innovation was not adopted at the moment of the collection of the data but there is some interest in trialing it again (mark=10%).

10) Style of Milking Shed and Milking System:

- The newest milking sheds and milking systems in *Farm 1* are very similar to those of New Zealand (for more details see section 6.2). In *Farm 1*, they considered the possibility of adopting rotary milking sheds but they had not done so yet because the rotaries are more expensive per cow milked, and because they think that in large dairy farms they would have too many problems with excess of mud around the milking shed.

- They are using New Zealand style milking sheds and systems because it costs less per cow milked per hour, and they provide enough comfort to the people who milk the cows.

- Even though, at the moment, they have adopted this innovation in only 2 or 3 of the dairy farms, they are convinced that this is the style that they prefer and that they would use in all new milking sheds (mark=80%).

Table 13: Summary of Innovations Adopted or Rejected by *Farm 1*

New Zealand Innovations	Mark
1) Focus on Production per Hectare	1
2) Give Marked Importance to Pasture Production	0.25
3) Quantitative Pasture Monitoring	0.25
4) Utilization of Formal Pasture Budgets	0
5) Skilled and Motivated People Working on Farms	0
6) Less than 15 cows per Set of Teat-cups	0.45
7) Seasonal Calving, One or Two Calving Periods per Year	0.17
8) New Zealand Genetics	0.2
9) Rearing of Calves in Groups	0.1
10) Style of Milking Shed and Milking System	0.8
TOTAL (out of 10)	**3.2**

7.1.2 Farm 2

General Characteristics

The Farm Business

Farm 2 is the dairy enterprise of a larger farm business; the complete farm business also comprises a cropping enterprise and two more dairy farms on leased land. The dairy enterprise started in 1984 and has been continually growing since then. *Farm 2* is known in the region as an example of the adoption of New Zealand principles to Argentina.

Farm 2 is considered a model for a very specific production system that is based on trying to achieve excellence in all aspects, and on the high motivation of the people working in it.

The Farmer

Farmer 2 is the main owner of the farm; he makes all the strategic decisions. He is also in charge of supervising the different enterprises and sections of the farm business, of training and giving advice to the people in charge of each enterprise or section (dairy farmers and others) of the farm business, and he is very involved in recruiting new staff.

As was mentioned before, *Farmer 2* is a renowned innovator within the dairy farm sector especially in the adoption of New Zealand principles to Argentina. He is a cosmopolitan person who reads scientific and farming journals from several areas of Argentina and other parts of the world. He became interested in the New Zealand systems when he started dairying. He travelled for the first time to New Zealand in 1992 and since then has been in contact with the country's dairy sector.

Farmer 2 is convinced that all New Zealand innovations proposed in the present study can be beneficial to Argentine dairy farmers and he believes that in the future Argentine dairy farmers would have to change their present production systems to be more similar to the New Zealand systems in order to remain competitive.

Feeding Sources

- 51% of total annual requirements for the cows (lactating and dry) are produced on the milking platform (grazed grass, and winter and summer crops).

Figure 55: Feeds Eaten by *Farm 2* Cows

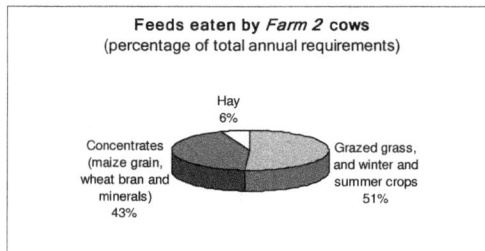

Feeds eaten by *Farm 2* cows
(percentage of total annual requirements)

Hay
6%

Concentrates
(maize grain,
wheat bran and
minerals)
43%

Grazed grass,
and winter and
summer crops
51%

- The company produces its own maize grain and buys wheat bran; the cropping enterprise sells the grains to the dairy enterprise at market prices.

- In *Farm 2*, silages are not made, maize is used in the form of grain and usually there is no excess of grass to be conserved (all grass is eaten directly by the cows).

New Zealand Innovations Adopted and/or Rejected

In *Farm 2*, not only was the owner interviewed but two couples that manage two dairy farms were also interviewed.

1)   Focus on Production per Hectare:

- In *Farm 2*, the milksolids production per hectare is their ultimate physical indicator. They monitor on a monthly basis a number of physical indicators for the different dairy farms of *Farm 2* and other dairy farms of the company, and they contrast them with indicators of other companies; one of those indicators is the production of milksolids per effective hectare.

- In *Farm 2*, they are focused on production per hectare because is their most limiting production factor (the farm's factors of production are land, labour and capital). With some effort they can find the right people for dairy farming and also operating capital, but land is usually expensive to buy and is difficult to lease for 7 to 10 years (approximate timeframe needed for a dairy business on leased land to pay off the investment and the effort).

- This innovation was completely adopted in *Farm 2* (mark=100%).

2)   Give Marked Importance to Pasture Production:

- *Farmer 2* mentioned that pasture production, pasture utilization and longevity of pastures are all important aspects for them. One of the dairy farmers of *Farm 2* stated that they know that the concentrates are expensive in comparison to pasture, consequently they try to produce and utilize as much pasture as they can.

132

- Another of the dairy farmers stated: *'we supplement with concentrates and most Argentine farmers supplement with grass. We try to get the most from our pastures and then we supplement with concentrates; that's why, even supplementing strongly, we consider that we have a pastoral system'.*

- In *Farm 2*, they are investing heavily in raising the level of phosphate in their soils to reach 20 parts per million (ppm) of phosphates extracted by Olsen recommended by some Argentine experts.

- The dairy farmers mentioned that they look after the pastures, they try not to let the cows graze too low and they protect the newest pastures on the wet days. In some very cold winters when there is no pasture growth, to prevent the pasture cover from decreasing too much, they feed the cows only with supplements.

- *Farmer 2* usually spends some time training new dairy farmers in how to graze the different grass species at different times of the season.

- *Farm 2* is very focused on pasture production (mark=100%).

3)   Quantitative Pasture Monitoring:

- *Farm 2* dairy farmers monitor pasture stock (instead of pasture cover) on a regular basis. They do visual assessment and include only the pasture that could be eaten by the milking cows. This is slightly different to New Zealand where all grass above soil level is included in the assessment. They believe that this way is more appropriate for their pastures and their production system. From the assessment that they do, they calculate the total stock of grass, as kilograms of dry matter of grass available for cows.

- Every new dairy farmer is trained by cutting grass samples in different paddocks, with different grass species and at different times of the year. Those samples are dried in a stove or microwave and then weighed.

- *Farmer 2* believes that in order to manage pasture efficiently, constant monitoring is essential. One of the dairy farmers mentioned that they monitor the pasture cover and growth frequently (especially at some times of the year) in order to regulate the quantity of concentrates that are fed.

- It can be considered that they adopted this New Zealand innovation with some small modifications (mark=90%).

4) Utilization of Formal Pasture Budgets:

- A feed budget is planned on a computer spreadsheet at the beginning of the year, and updated periodically with the actual stocks of dry matter of grass and other feeds. Decisions to buy feedstuffs and to fertilise are based on the feed budget.

- Some years ago only *Farmer 2* used to do the feed budgets of the three dairy farms. Nowadays he is still doing a centralized feed budget and some of the dairy farmers do theirs as well.

- Every year *Farmer 2* and the dairy farmers have to take an important decision at the beginning of the spring as to when they decide to stop feeding supplements because there is enough pasture and pasture growth to cover the cows' requirements.

- *Farmer 2* plans and monitors frequently a feed budget for all the dairy farmers in order to help them to take the decisions to buy foodstuffs or to fertilize with nitrogen. Other reasons are to keep records of stock of pasture and cows intake at different times of the year and through several years, and consequently to increase the prediction accuracy each year. The pasture budget also helps him to find problems in pasture production or pasture utilization on the different dairy farms.

- This innovation was completely adopted by *Farm 2* (mark=100%).

5) Skilled and Motivated People Working on Farms:

- The three dairy farms of *Farm 2* are managed separately. The operating structure of each dairy enterprise is quite similar to a New Zealand "share-milking" contract (for more details see LIC, 2003/04). The two parts that sign the contract are "the dairy farmers" (usually a young couple) and "the administrator" (in the three farms of the present study the administrator is *Farmer 2*). Both parts receive wages from the dairy enterprise per day worked. The dairy enterprise leases all the land and the cows. All the costs and returns of the dairy enterprise are shared equally between the two parts. In the three dairy enterprises of *Farm 2* most of the land and cows is owned by *Farmer 2*.

- The dairy farmers in *Farm 2* are all university graduates most of them from careers related to agriculture. When they started working in *Farm 2* they were trained in one of the dairy enterprises, and some of them became the dairy farmers after the training period. The dairy farmers are responsible for most of the decisions within their dairy enterprise and at the same time are very involved in the daily duties of the farm. One member of the couple is usually in charge of one of the daily milkings, they usually milk at 6 a.m. and at 3 p.m. (not very common in Argentina).

- This is the New Zealand innovation that is considered as the most important for *Farmer 2*. He believes that all the other innovations are adopted in order to be able to get skilled and motivated people working in dairy farm businesses. He thinks that with the typical Argentine system it is not possible to provide the working conditions and enough motivation required by people of this kind.

- He mentioned in the interview that this was the aspect that impressed him the most in his first visit to New Zealand. He was surprised that the farm owner could delegate most of the dairy farm activities to other people, and then have the opportunity to focus on other activities.

- *Farm 2* has completely adopted this New Zealand principle of having skilled and motivated people in charge and working on the dairy farms (mark=100%).

6) Less than 15 cows per Set of Teat-cups, and Other Innovations that Impact on Labour Productivity:

- In *Farm 2*, there are approximately 25 cows milked per set of teat-cups. They have a policy that nobody should milk more than four hours per day. Consequently, they implemented milking shifts in order to prevent any person from milking more than 4 hours per day. They are aware that they have too many cows for the milking sheds that they have. *Farmer 2* stated that they are thinking about once a day milking in order to have the possibility to increase the cow numbers without the need to invest in a new milking shed, while continuing to limit the time spent milking by each person to 4 hours per day.

- One of the business goals in *Farm 2* is to have high labour productivity in order to be able to employ less people that are better paid individually. Therefore, they are continually investing in plant, machinery and buildings in order to meet this goal.

- Other technologies used in *Farm 2* are: motorbikes, front loaders on tractors, automatic effluent spreaders and motorised backing-gates.

- In *Farm 2*, this innovation was partially adopted (mark=50%).

7)   Seasonal Calving, One or Two Calving Periods per Year:

- Before converting to dairy they were beef farmers, and produced their own calves. The beef cows in Argentina typically calve seasonally. From the beginning of the dairy enterprise they planned to continue with a seasonal system but with dairy cows.

- In *Farm 2*, they decided to adopt seasonal calving in order to provide better working conditions for the dairy farmers. They think that seasonal calving gives the people on the dairy farms the opportunity to focus on one important task at one time. They can, for example, plan the mating, train the people and themselves for it, then execute the plan, and finally measure the results. It is also an advantage for them to have high workload periods alternated with less intensive periods.

- *Farmer 2* also mentioned the following advantages: the need for pregnancy testing only twice a year (in all-year round calving dairy farms, pregnancy testing is done periodically), the possibility to raise the replacement heifers in a more simple and orderly fashion because they are all of the same age, have some periods without calving and mating. *Farmer 2* thinks that everything is more simple if it is seasonal.

- It is important to mention that from the four dairy farmers interviewed (apart from *Farmer 2* who is the administrator) three of them agreed in ranking seasonal calving as one of the most important New Zealand innovations. Most of them would not work on an all year round calving dairy farm.

- *Farm 2* has recently started two other dairy farms in a different region of the country, also seasonal but in these cases with spring calving. They completely adopted this New Zealand innovation (mark=100%).

8)   New Zealand Genetics:

- In *Farm 2*, they use only New Zealand genetics. In the first years of the dairy enterprise they started buying some Jersey bulls from *Farmer 4*, and when they had the opportunity, they started to buy New Zealand semen.

- Up to the last season they have been breeding to have crossbreed cows; they have been putting New Zealand Jersey semen to those cows with a higher proportion of Holstein or Holstein Friesian, and New Zealand Holstein Friesian semen to those cows with a higher proportion of Jersey. Now they only use New Zealand Jersey semen.

- *Farmer 2* mentioned that they decided to use New Zealand genetics because, from the very beginning, they wanted to have a seasonal and pastoral dairy system, and New Zealand cows were selected for that kind of system.

- In *Farm 2* they believe that New Zealand genetics are beneficial for their systems. The four dairy farmers interviewed agreed in ranking the New Zealand Genetics as one of the three most important New Zealand innovations adopted by *Farm 2*. One of them, a veterinarian with some experience in Argentine Holstein cows, stated: *'these cows are machines for getting in calf. We found that the cows that were getting in calf for the second time had the highest in-calf rate of all. This is completely the opposite of what we learnt in the university and what usually occurs with Argentine Holstein cows. We think that this could be caused by the very strict selection for fertility that occurs in New Zealand herds.'*

- A disadvantage of New Zealand genetics is that the bull calves, especially the Jerseys, are very difficult to sell in the Argentine market. And that there is no market for cows and heifers of New Zealand genetics.

- They completely adopted this New Zealand technology (mark=100%).

9)   Rearing of Calves in Groups:

-   All the calves are reared in groups in the three dairy enterprises. Usually the
    woman of the dairy farming couple is in charge of rearing the calves.

-   In one of the dairy farms, the calves are fed with a round calf feeder that can be
    towed with a 4-wheel bike. The other two dairy farms use a 200 litre plastic
    barrel with approximately 20 "calf-teats" which is filled from a big tank attached
    to a trailer that is towed by a tractor.

-   The two women interviewed who are in charge of raising the replacement
    heifers in two of the dairy farms are convinced of the benefits of this New
    Zealand practice. However, they also agree that this could not be the best system
    for everyone. This calf rearing system requires that the person in charge must
    treat each calve individually even though they are in groups. Therefore this
    system could be beneficial for people interested in the calves and with the
    capacity to differentiate them individually within the groups.

-   In *Farm 2*, they have completely adopted this New Zealand practise
    (mark=100%).

10)   Style of Milking Shed and Milking System:

-   The first milking sheds that were built in *Farm 2* from 1984 to 1987 had two
    walls with two lines of pipelines and sets of teat-cups, very similar to most
    typical Argentine milking sheds.

-   However after the visit of *Farmer 2* to New Zealand in 1992 and with the
    increase in cow numbers, they had first to add more sets of teat-cups, and then to
    demolish the sheds and build new ones. The new milking sheds were built in the
    New Zealand way; they are one wall herringbone milking sheds, with typical
    New Zealand milking systems.

-   In *Farm 2*, they adopted these New Zealand style milking sheds and milking
    systems for two main reasons; firstly because they are designed to milk

maximum number of cows with the least time and labour, and secondly because of the relatively low cost of the structures.

- The new dairy sheds that were built in the two new dairy enterprises of *Farm 2* are also typical New Zealand sheds with single line swing over herringbone milking systems. *Farmer 2* thinks that the next milking shed could be a rotary milking shed and system. They have adopted completely this New Zealand technology (mark=100%).

Table 14: Summary of Innovations Adopted or Rejected by *Farm 2*

New Zealand Innovations	Mark
1) Focus on Production per Hectare	1
2) Give Marked Importance to Pasture Production	1
3) Quantitative Pasture Monitoring	0.9
4) Utilization of Formal Pasture Budgets	1
5) Skilled and Motivated People Working on Farms	1
6) Less than 15 cows per Set of Teat-cups	0.5
7) Seasonal Calving, One or Two Calving Periods per Year	1
8) New Zealand Genetics	1
9) Rearing of Calves in Groups	1
10) Style of Milking Shed and Milking System	1
TOTAL (out of 10)	**9.4**

7.1.3 Farm 3

General Characteristics

The Farm Business

Farm 3 is a dairy enterprise of a larger farm business; the complete farm business has farms at two sites approximately 300 km apart. In both sites, there is a dairy and a cropping enterprise. *Farm 3* is the dairy enterprise of one of the sites. The company is a family business owned and directed by several members of a family.

The owners of *Farm 3* are focused on their return on assets (ROA), and considering that their main asset is the land, they are particularly interested in the return per hectare. Consequently they add or subtract hectares to the dairy enterprise depending on how profitable this activity is in comparison to cropping.

139

The Farmer

Farmer 3 is one the owners of the farm business and the member of the family that is more involved in the managing of the farms. He is involved in most strategic and tactical decisions that are related to the production systems (in the dairy and cropping enterprises).

As was mentioned before (in section 3.1.4), *Farmer 3* is considered by others to adopt new technologies and ideas only when they have proven to be beneficial (from interviews with *Farmer 2* and *Farmer 4*). This farm is among the top performers of its discussion group (interviews with *Farmer 2* and *Farmer 4*). Due to his prestige and conservative views, *Farmer 3* can be considered to be an opinion leader within the farmers of his discussion group (see section 2.1.3.4 for definition of opinion leadership). Even though *Farmer 3* is considered to be sceptical about new technologies, he has adopted some NZ innovations.

Farmer 3 is a cosmopolitan person that reads local and foreign technical magazines. He has travelled to the United States and to New Zealand in order to understand other farm systems.

Feeding Sources

- 64% of total annual requirements for all dairy animals (milking and dry cows, calves and heifers) are produced on the milking platform (grazed grass, and winter and summer crops).

Figure 56: Feeds Eaten by *Farm 3* Dairy Animals

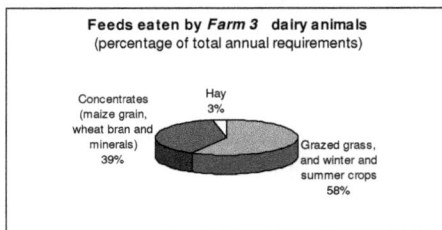

Feeds eaten by *Farm 3* dairy animals
(percentage of total annual requirements)

Concentrates (maize grain, wheat bran and minerals) 39%

Hay 3%

Grazed grass, and winter and summer crops 58%

- The company produces its own maize grain and buys wheat bran; the cropping enterprise sells the grains to the dairy enterprise at market prices.

- In *Farm 3* (as in *Farm 2*) silages are not made, maize is used in the form of grain. Hay is usually produced in the farm from a summer crop called "Moha" (*Setaria Italica*).

New Zealand Innovations Adopted and/or Rejected

1)   Focus on Production per Hectare:

- In *Farm 3* they try the get the highest possible productivity per effective hectare, at the same time they monitor the ratio between the prices of concentrates and milk in order to increase or decrease the level of supplementation. For example, they know that they should not spend, during winter, more than 40% (approximately) of the gross milk income on concentrates.

- Typically, large Argentine dairy farms also have a cropping enterprise and can easily change hectares between the two enterprises, this introduces a more complex situation when the optimum stocking rate is being decided.

- In *Farm 3* they monitor more frequently the production per cow than the production per hectare, because once the stocking rate and the level of supplements are decided, the production per hectare depends mostly on the productivity per cow.

- In *Farm 3* they are very clear that, within the physical indicators, production per hectare is the most important, this approach is very similar to that of New Zealand farmers (mark=100%).

2)   Give Marked Importance to Pasture Production:

- In *Farm 3*, they are slowly increasing the phosphate levels of the soils; they are trying to have 20 ppm of phosphates extracted by Bray and Kurtz (or Bray 1). Some experts suggested to *Farmer 3* that he should increase the phosphate levels up to 25 ppm for lucerne and up to 15 ppm for fescue.

- In *Farm 3*, when they have to choose between pastures and cows they usually give priority to cows, therefore in some periods of the year in which cows' requirements are much higher than pastures growth, they usually graze the pastures very intensively leaving lower residuals than those recommended by experts.

- Increasingly in *Farm 3*, they are giving more importance to pasture production. *Farmer 3* is not completely convinced that it is beneficial to fertilize or to look after their pastures. However, he is slowly becoming convinced as some consultants and other top farmers are doing it.

- In *Farm 3* they have adopted this New Zealand principle partially (mark=60%).

3)   Quantitative Pasture Monitoring:

- Approximately 2 years ago, in *Farm 3*, they adopted a programme to record pasture cover and growth. One person is in charge of going around all paddocks of the three dairy farms once a week, or once every two weeks, depending on the time of the year. The results of each monitoring are reported to *Farmer 3*, to the person in charge of the whole dairy enterprise and to each dairy farmer.

- *Farmer 3* finds that the adoption of this innovation has several advantages: it helps them to better manage the cows' supplementation; it gives them useful information about the stock of grass available to be eaten by the cows (they calculate a ratio of the kilograms of dry matter available per milking cow); in some circumstances they use the data to redistribute hectares across the dairy farms when necessary; and it helps them to have a minimum stock of dry matter of grass on each dairy farm.

- In *Farm 3* they adopted this practice in a similar way as is done in New Zealand. The main difference from a typical New Zealand farm would be that in New Zealand, the dairy farmer (who milks the cows and decides the paddock that the cows will graze) does the pasture monitoring. This difference could be quite significant because information and time are lost in the process of reporting the results of the monitoring to the dairy farmer (mark=70%).

4)   Utilization of Formal Pasture Budgets:

- In *Farm 3*, feed budgets are utilized in which are included pastoral budgets. Pasture growth was measured during several years in the nineties, and they use average monthly figures for their pastoral budgets. They usually do a pasture budget for the year, not only for the milking cows but also for all the animal categories in all the farms. During the year, these pasture budgets are checked three or four times and modifications are done if necessary.

- *Farmer 3* uses pasture budgets in order to decide stocking rates and level of supplementation needed. He finds pasture budgets useful to decide the amount of supplements that they need to buy.

- In *Farm 3*, they adopted this technique completely and in the same way as it is used in some New Zealand farms. However, in comparison to what was observed in *Farm 2* the pasture budget is monitored less frequently in *Farm 3* (mark=90%).

5)   Skilled and Motivated People Working on Farms:

- In addition to *Farmer 3*, who spends approximately half of his time working for the dairy enterprise, there are two more people involved in the "management team" of the three dairy farms. One of them lives near the dairy farms and spends most of his time on the dairy enterprise (from now on he is called the "dairy enterprise supervisor"). The other person is a veterinarian who visits the dairy farms one or twice a week. In addition, there is a dairy farmer in charge of operating each of the dairy farms. Some of the members of the dairy farmers' family usually work with them.

- *Farmer 3*, the dairy enterprise supervisor, and the veterinarian are all university graduates. The dairy farmers have finished primary or secondary school.

- *Farmer 3* mentioned that he understands some of the advantages of having more skilled and motivated people, however he is not sure whether the benefits are higher than the costs. He finds the transition from typical Argentine dairy farmers to more skilled and motivated people to be very costly. A big investment would have to be made to enlarge the milking sheds and to repair the dairy

143

farmers' houses. However, the biggest cost would be the training of those new dairy farmers. He thinks that at the moment these new people would need to be trained directly by him. He stated that the dairy farms' performance would have to be much better in order to compensate for the investment and the higher income that this new kind of dairy farmers should receive.

- The working environment, the job conditions and the level of education of the people working on the dairy farms of *Farm 3* were quite typical of Argentina. Therefore they have not adopted this New Zealand principle (mark=0%).

6) Less than 15 cows per Set of Teat-cups, and Other Innovations that Impact on Labour Productivity:

- In *Farm 3* there are approximately 26 cows milked per set of teat-cups (1.73 times more cows per set of teat-cups than in New Zealand).

- *Farmer 3* believes that to improve labour productivity they would need more skilled and motivated people and at the same time more investment in plant, machinery and houses. He thinks that the dairy farmers who are actually working in *Farm 3* dairy farms would not use efficiently the spare time given by, for example, larger milking sheds. Consequently, investment in plant, machinery and buildings has to go together with the employment of more skilled and motivated people.

- In *Farm 3* this New Zealand innovation was adopted only in a small proportion (mark=30%).

7) Seasonal Calving, One or Two Calving Periods per Year:

- The three dairy farms of *Farm 3* have cows calving during ten months of the year (they try not to have cows calving in January and February).

- *Farmer 3* mentioned two main reasons for rejecting seasonal calving. The first reason is the seasonality of the milk price in Argentina; usually the highest prices are paid during winter, but he thinks that is not always the case and consequently is quite risky to have a big proportion of milk produced in a short period of time. The second reason is that they cannot get all their cows in calved each 12 moths; on average their cows need 13 months between two calvings, some of those cows are very high producers with long lactations.

- In *Farm 3,* they have not adopted this New Zealand innovation (mark=0%).

8)   New Zealand Genetics:

- In *Farm 3*, they use Argentine Holstein genetics for all their cows and heifers. They use Argentine Holstein bulls with negative values for live weight with the intention to moderate the cows' size; their adult cows weigh, on average, 500 kilograms instead of more than 550 kilograms of a normal Argentine Holstein cow.

- In other dairies of the company, they use a small proportion of New Zealand genetics (mostly Holstein Friesian).

- Firstly they prefer Argentine Holstein because the North American genetics (which have a big influence in Argentine Holstein cows) are selected from a bigger population and have been selected for a larger number of years. Secondly, because they are paid mainly for milk-protein and New Zealand cows have higher proportion of milk-fat than the Argentine Holstein cows, considering also that milk-fat is energetically more expensive to produce than milk-protein. And thirdly, they prefer Argentine Holstein because is cheaper and easier to find than New Zealand semen.

- In *Farm 3* they have not adopted this New Zealand technology (mark=0%).

9)   Rearing of Calves in Groups:

- In *Farm 3*, they adopted this New Zealand innovation for raising most of the calves of the three dairy farms. They adopted this innovation previously in other dairy farms of the company, and approximately 2 years ago they adopted in *Farm 3* because it increases labour productivity (mark=100%).

10)   Style of Milking Shed and Milking System:

- Only the newest of the three milking sheds was built in the New Zealand style. The other two milking sheds are typical Argentine milking sheds. Two of the milking systems are high-line single herringbone parlours (also called "swing over" parlours) and one has two low pipelines with sets of teat-cups. *Farmer 3* mentioned that they are planning to change the milking system of this last dairy farm to the New Zealand style.

- In *Farm 3*, they have chosen New Zealand style milking sheds mainly because of their lower cost, and they adopted New Zealand milking systems because more cows can be milked per set of teat-cups than in other milking systems.

- In conclusion in *Farm 3* they are changing to milking sheds and systems that are common in New Zealand, and they are convinced that is the style that they would use in future new milking sheds (mark=100%).

Table 15: Summary of Innovations Adopted or Rejected by *Farm 3*

New Zealand Innovations	Mark
1) Focus on Production per Hectare	1
2) Give Marked Importance to Pasture Production	0.6
3) Quantitative Pasture Monitoring	0.8
4) Utilization of Formal Pasture Budgets	0.9
5) Skilled and Motivated People Working on Farms	0
6) Less than 15 cows per Set of Teat-cups	0.3
7) Seasonal Calving, One or Two Calving Periods per Year	0
8) New Zealand Genetics	0
9) Rearing of Calves in Groups	1
10) Style of Milking Shed and Milking System	1
TOTAL (out of 10)	**5.6**

7.1.4 Farm 4

General Characteristics

The Farm Business

Farm 4 is the largest dairy farm of a family farm business that is mainly focused on dairy farming. This farm business breeds and commercialises New Zealand Jersey heifers and bulls.

The mission of *Farm 4* is the following: *'look for excellence within farming, for the satisfaction of all the members that work on the company, and for the good of the community. Contribute to improve the image of the Argentine dairy sector'*. Probably the main characteristic of *Farm 4* is a production system completely consistent with the goals and values of its owners.

The Farmer

Farmer 4 is one the owners of the farm business and member of the family in charge of its direction. He is responsible for taking most of the strategic decisions of the farm business. He lives on the farm and is involved in the daily running in the business.

As was previously mentioned, *Farmer 4* was one of the first Argentine farmers to adopt New Zealand innovations and one of the most convinced of the benefits of the New Zealand dairy farm principles. *Farmer 4* is a cosmopolitan person who is both aware of the evolution of the Argentine dairy sector and of the changes that occur in the international dairy markets.

Farmer 4 is in close contact with New Zealand and Australian dairy production sectors. *Farmer 4* first travelled to New Zealand in 1982, initially interested in the genetics of New Zealand dairy cows.

Feeding Sources

- 45% of total annual requirements for all dairy animals (milking and dry cows, calves and heifers) are produced on the milking platform (grazed grass, and winter and summer crops).

Figure 57: Feeds Eaten by *Farm 4* Dairy Animals

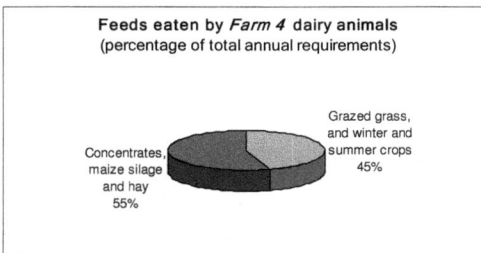

Feeds eaten by *Farm 4* dairy animals
(percentage of total annual requirements)

Grazed grass, and winter and summer crops 45%

Concentrates, maize silage and hay 55%

- The company produces maize silage, maize grain and buys wheat bran. The
  cropping enterprise sells the grains and silage to the dairy enterprise at market
  prices.

New Zealand Innovations Adopted and/or Rejected

1) Focus on Production per Hectare:

- In *Farm 4* milksolids production per hectare is their ultimate physical indicator,
  because this helps the business to meet their financial goals.

- This innovation was completely adopted in *Farm 4* (mark=100%).

2) Give Marked Importance to Pasture Production:

- *Farmer 4* and the team that manages *Farm 4* believe that their knowledge of
  their pastures is one of the strengths of their system.

- In *Farm 4* pastures are fertilized strongly, at higher levels that on most
  Argentine dairy farms. Levels of 25 ppm of phosphate (Olsen), 15 ppm of
  sulphates are maintained, 150 units of Nitrogen are applied per hectare each
  season. The target in *Farm 4* is to produce 12 tonnes of dry matter of grass per
  hectare per year in their best soils.

- *Farmer 4* has calculated that in his production system the dry matter from
  grazed grass is the cheapest feed for their cows. Therefore he tries to produce as
  much grass dry matter, as is economically feasible. However they are also
  aware that in Argentina supplements are relatively cheap and they utilize them
  as long as it is convenient for them.

- In *Farm 4* this New Zealand principle is completely adopted (mark=100%).

3) Quantitative Pasture Monitoring:

- Pasture cover is estimated on a regular basis on *Farm 4*. The person in charge
  uses a rising plate meter and also visual assessment. Based on this information
  he calculates the average pasture cover and the pasture growth rates. Then,

together with the dairy farmer, they decide the pastures to be grazed and the pre and post grazing covers wanted.

- In *Farm 4* quantitative pasture monitoring is nearly completely adopted. The only deviation from the typical New Zealand farm is that it is not the dairy farmer who monitors the pasture-cover (mark=90%).

4) Utilization of Formal Pasture Budgets:

- A pastoral budget is planned on a computer spreadsheet at the beginning of the year and updated periodically with the actual pasture covers and growths. The pastoral budget is an essential tool for grazing management for *Farm 4*.

- In *Farm 4* pastoral and feed budgets are tools that help in the making of medium-term decisions on feeding. The budgets help to assess the need to buy or make more supplements or the need to fertilize pastures with nitrogen.

- This innovation was completely adopted by *Farm 4* (mark=100%).

5) Skilled and Motivated People Working on Farms:

- The management team of *Farm 4* is comprised of *Farmer 4* ("the managing director"), one person in charge of the dairy enterprise ("the dairy manager"), one person in charge of the machinery enterprise, the cropping and the rearing of replacements ("the cropping manager"), and a person in charge of the administrative tasks ("financial manager").

- There are two families of dairy farmers in charge of the dairy farm operational activities. One family is in charge of the autumn calving herd and the other is in charge of the spring calving herd. There is a third group of four people, who help the two families of dairy farmers at the times of greatest activity (calving and mating).

- The people in charge of the dairy farm operational activities have finished primary or secondary school. The dairy manager, who is a university graduate, is very involved in the dairy farm and works closely with the dairy farmers.

- In *Farm 4* they adopted this innovation because they believe that, in order to be efficient in dairy farming, the people involved have to be trained and have to be motivated.

- In *Farm 4* people that typically work in Argentine dairy farms are employed to operate the dairy farm. What is innovative on *Farm 4* is that they try to select the best people and they continue training them. Another innovative practice is the fact that the owner of the farm and other skilful people (in technical matters) work closely with the dairy farmers.

- *Farmer 4* mentioned that he believes that an operating structure similar to the share-milking contract in New Zealand is possible in Argentina. However he believes that *Farm 4* is too large and complicated to be operated within that structure.

- The job conditions, working environment and the level of education of the people in charge of the dairy farm operational tasks are higher than those of typical Argentine dairy farms. However they have not completely adopted this New Zealand innovation (mark=80%).

6) Less than 15 cows per Set of Teat-cups, and Other Innovations that Impact on Labour Productivity:

- In *Farm 4* approximately 24 cows are milked per set of teat-cups (1.6 times more cows per set of teat-cups than in New Zealand) (mark=40%).

- In *Farm 4* this New Zealand principle was only partially adopted because they believe that in Argentina the economic conditions are different to New Zealand, and that given the local environment, it is probably more convenient to have more people working on the farm and less investment in plant and machinery.

7) Seasonal Calving, One or Two Calving Periods per Year:

- *Farm 4* has two seasonal calving periods per year, one in autumn and one in spring.

- *Farmer 4* has completely adopted this New Zealand innovation (mark=100%).

8)   New Zealand Genetics:

-    In *Farm 4*, only New Zealand genetics is used, and nearly all the cows are Jerseys. *Farmer 4* is known for having high genetic merit Jersey cows and for selling Jersey bulls and heifers.

-    In *Farm 4*, this New Zealand technology was completely adopted (mark=100%).

9)   Rearing of Calves in Groups:

-    All calves are reared in groups. They completely adopted this New Zealand practise (mark=100%).

10)   Style of Milking Shed and Milking System:

-    *Farm 4* is one of the few dairy farms in Argentina with a rotary milking shed and milking system. The design and most of the parts were brought from New Zealand. This innovation is completely adopted by *Farm 4* (mark=100%).

Table 16: Summary of Innovations Adopted or Rejected by *Farm 4*

New Zealand Innovations	Mark
1) Focus on Production per Hectare	1
2) Give Marked Importance to Pasture Production	0.8
3) Quantitative Pasture Monitoring	0.8
4) Utilization of Formal Pasture Budgets	0.9
5) Skilled and Motivated People Working on Farms	0.8
6) Less than 15 cows per Set of Teat-cups	0.4
7) Seasonal Calving, One or Two Calving Periods per Year	1
8) New Zealand Genetics	1
9) Rearing of Calves in Groups	1
10) Style of Milking Shed and Milking System	1
TOTAL (out of 10)	**9.1**

7.1.5 Farm 5

General Characteristics

The Farm Business

A single family owns *Farm 5*. It is directed by one of its owners *(Farmer 5),* and at least one additional member of the family works for the business. *Farm 5* is a farm business without land, as all the land is leased and most of the plant, machinery and cows are the assets of the business. There are not many dairy farmers in Argentina that lease all the land (researcher's experience in Argentine dairy production sector). In New Zealand dairy farms on leased land are also rare (IFCN, 2002).

The family is also involved in other businesses, some of them related to farming. They are focused on maximizing returns on capital and they are interested in being diversified in order to reduce risks. *Farmer 5* mentioned that they are not interested in having their capital tied to physical assets, but they prefer to keep a high level of liquidity in the business in order to take advantage of financial opportunities.

The Farmer

Farmer 5 is the main decision maker of the business; he is very involved in all strategic decisions and in the main tactical decisions.

Farmer 5 is a very entrepreneurial person, he also owns (at least in part) and directs a business that sells, repairs and maintains milking machines, milk tanks and cooling systems. He has also worked as a dairy farm consultant in the past.

Farmer 5 is very informed about the evolution of the Argentine dairy sector, reads most relevant Argentine literature about dairy farming and is aware of the principles of dairy farming in other countries like the Unites States of America, New Zealand and Australia. However, he can be considered to be more locally oriented than *Farmers 1, 2, 3* and *4*.

Feeding Sources

- In *Farm 5* in the last season (2003/04) approximately 70% of the total requirements of cows, heifers and calves were brought from outside the milk platform (maize silage and concentrates).

152

Figure 58: Feeds Eaten by *Farm 5* Dairy Animals

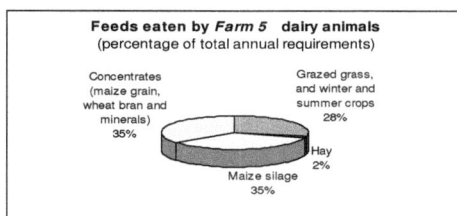

Feeds eaten by *Farm 5* dairy animals
(percentage of total annual requirements)

Concentrates (maize grain, wheat bran and minerals) 35%

Grazed grass, and winter and summer crops 28%

Hay 2%

Maize silage 35%

- Maize silage is used abundantly, especially during the winter. It is fed by a system that it is called "self-feeding". The cows eat by themselves from the big plastic bags where the silage is stored. They can eat approximately 20 kilograms of silage (green matter) in approximately two hours. *Farmer 5* said that the wastage is very low.

New Zealand Innovations Adopted and/or Rejected

1) Focus on Production per Hectare:

- The fact that all the land is leased puts a lot of pressure on the returns per hectare. Consequently, *Farm 5* is very focused on maximizing production per hectare provided that the increase in productivity has a positive impact on the financial returns per hectare.

- In *Farm 5* this focus is completely adopted (mark=100%).

2) Give Marked Importance to Pasture Production:

- Farmer 5 clearly stated that he gives priority to the cows over the pastures. He said 'at the moment the cow is more important than pastures. For me, cows are sacred, like in India, I try to feed them as well as possible even if I have to sacrifice the pasture'. Consequently, very often the cows graze the pastures too low or go into a paddock despite it being too wet.

- However in Farm 5, they give a lot of importance to pasture and crop productivity because (as was mentioned before) the rent of the land is one of the highest costs. They have noticed that on their land, the pastures produce much

153

more in the first three years of production, therefore Farmer 5 decided to replant pastures every three years, instead of the traditional 4 years, in order to eliminate the fourth year of low pasture productivity.

- In Farm 5 they decided not to invest in phosphate fertilization because they do not know if the landowner would renew the leasing contract. They fertilise with small quantities of phosphate and nitrogen at the moment of replanting.

- Farmer 5 says that he enjoys grazing cows, and that he would give more priority to pastures if he had the time or if he owned the land (he would fertilize more). He also emphasised that only 28% of the requirements of the cows, heifers and calves are covered by grazed grass and therefore the focus is on utilizing silage, hay and concentrates (which cover 70% of all the requirements).

- In Farm 5 this principle was not adopted (mark=0%).

3)   Quantitative Pasture Monitoring:

- In *Farm 5*, all the grass available is utilized, and for the reasons explained in the previous point, pasture is not considered of sufficient importance to justify more time or resources spent on this New Zealand practice.

- In *Farm 5* this principle was not adopted (mark=0%).

4)   Utilization of Formal Pasture Budgets:

- In *Farm 5* this principle was not adopted (mark=0%).

5)   Skilled and Motivated People Working on Farms:

- *Farmer 5* is a national university graduate from a career called "agricultural engineering". He plans and supervises most of the activities of the farm.

- People that work on the farm are all employees; all of them receive wages for their work. He mentioned that this is quite uncommon in the region where most of the people in charge of a dairy farm receive from 8 to 9% of the milk proceedings.

- All employees have finished primary or secondary school without any other formal training in farming. However *Farmer 5* makes the effort of supervising them and training them constantly.

- In *Farm 5* people from the region are employed. These people are the people that traditionally work in dairy farming in Argentina. *Farmer 5* mentioned that he has tried to motivate and train the people on his farm and he believes that he found a limit. After reaching a certain level of income and conditions of living and working, it seems that staff is not motivated by further improvements. He thinks that in order to get returns from additional incentives and improvements of working and living conditions, he would have to look for people outside the traditional dairy-farming environment (for example young graduates from technical schools or universities). However, he is uncertain if that kind of people would really compensate for the additional costs and investment with more efficiency.

- The working and job conditions are slightly above those of typical Argentine dairy farms, and the farmer trains the staff on the job. This New Zealand principle is partially adopted (mark=30%).

6)   Less than 15 cows per Set of Teat-cups, and Other Innovations that Impact on Labour Productivity:

- In *Farm 5* there are approximately 30 cows milked per set of teat-cups, with two shifts of people to milk the cows (2 times more cows per set of teat cups than in New Zealand).

- In *Farm 5* they achieve a relatively high number of cows per person based on an extreme simplification of the system, and a high motivation to perform from the relatively high wages and close supervision.

- *Farmer 5* believes that the ratio between investment in plant, machinery and buildings and the number of people working on the farm is well balanced for his production system. It is important to take into account that *Farm 5* is in leased land and that this limits the intention to invest in its overall structure.

- The New Zealand innovation principle of investing more in plant, machinery and buildings in order to maximize labour productivity was not adopted (mark=0%).

7)   Seasonal Calving, One or Two Calving Periods per Year:

- In *Farm 5* there are two calving periods, one in autumn (February, March and April) and one in spring (July, August and September).

- *Farmer 5* explained that the calving period in autumn is to take advantage of the high milk prices during the winter; and that the spring calving is for financial reasons because they have to continue paying expenses during the summer. In addition, because during the first months of the year there is usually a slight increase in the milk price.

- In *Farm 5,* this New Zealand innovation was completely adopted (mark=100%).

8)   New Zealand Genetics:

- *Farmer 5* is implementing a breeding scheme with three breeds: Argentine Holstein, New Zealand Holstein Friesian and New Zealand Jersey. He is convinced that New Zealand genetics are adequate for his system however he wants to take the most from the heterosis.

- All heifers in *Farm 5* are inseminated with New Zealand Jersey. The Argentine Holstein semen utilized is with negative values for live weight and for calving difficulty. Target average adult live weight of the herd is between 450 and 470 kilograms.

- *Farmer 5* stated: *'The possibility of having a seasonal calving system is given by the breed, as you can not have a seasonal calving with Argentine Holstein cows because the minimum interval between two calvings of a Holstein cow is around 14 months'. 'Seasonality is related to fertility and fertility is related to type of cow'. Farmer 5* uses New Zealand genetics mainly for robustness and fertility.

- Approximately two thirds of *Farm 5's* herd is New Zealand Genetics and one third are Argentine Holstein. Therefore this New Zealand technology was

156

partially adopted in *Farm 5* (mark=67%). However, it is the opinion of the researcher that Argentine genetics are used for clear specific reasons.

9)   Rearing of Calves in Groups:

- They completely adopted this New Zealand practice for autumn and spring calves (mark=100%).

10)   Style of Milking Shed and Milking System:

- In *Farm 5* the milking shed and milking system are New Zealand style because they are of relatively low cost and simple (mark=100%).

Table 17: Summary of Innovations Adopted or Rejected by *Farm 5*

New Zealand Innovations	Mark
1) Focus on Production per Hectare	1
2) Give Marked Importance to Pasture Production	0
3) Quantitative Pasture Monitoring	0
4) Utilization of Formal Pasture Budgets	0
5) Skilled and Motivated People Working on Farms	0.3
6) Less than 15 cows per Set of Teat-cups	0
7) Seasonal Calving, One or Two Calving Periods per Year	1
8) New Zealand Genetics	0.67
9) Rearing of Calves in Groups	1
10) Style of Milking Shed and Milking System	1
TOTAL (out of 10)	**5**

7.1.6 Farm 6

General Characteristics

The Farm Business

Farm 6 is a family business; several members of a family owning it. The father *(Farmer 6)* and a son work full time for the farm business. Other family members help on the farm sporadically. The father was slowly passing the direction of the business to the son.

The business started as a family project. The farm was bought and slowly they have been developing it for irrigation, through flood-irrigation. The family started with the dairy enterprise six years ago and they have been investing all the profit in continuing to develop the farm and increase the cow numbers.

The farm is situated in a remote area of the country, far from populated areas. There are not many dairy farms close by and the dairy company that buys the milk is small and is situated relatively far from the farm. Additionally, to reach the farm it is necessary to drive 35 kilometres on a dirt road that is in bad condition.

Farmer 6 mentioned: *'...we wanted to develop a low cost dairy production system. We have the goal of being the last dairy farm in the country to close if necessary...'*

The Farmer

Farmer 6 is a very experienced person in dairy production systems. He has worked as a dairy farm consultant for many years in very different countries and conditions and he mentioned that he tried the *'New Zealand dairy production principles'* in all of them with very good results.

Feeding Sources

- In *Farm 6* the farmers do not carry quantitative records, therefore only general data about the system is detailed in this section.

Figure 59: Feeds Eaten by *Farm 6* Lactating Cows

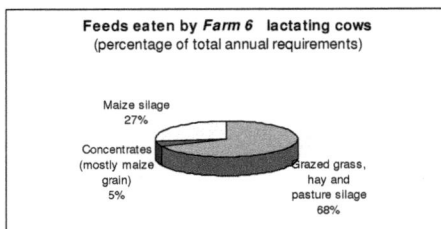

Feeds eaten by *Farm 6* lactating cows
(percentage of total annual requirements)

Maize silage
27%

Concentrates
(mostly maize
grain)
5%

Grazed grass,
hay and
pasture silage
68%

- Nearly all of the feeds are produced on the farm.

New Zealand Innovations Adopted and/or Rejected

1) Focus on Production per Hectare:

-   *Farm 6* is a self-contained farm in which the maize silage and the concentrates are produced on the farm. Their target at the moment is to produce approximately 15,000 litres of milk per hectare (approximately 1,080 kilograms of milksolids per hectare).

-   *Farmer 6* stated they are more focused on maximizing production per hectare than production per cow. He also mentioned that they make the cows graze too low for maximizing production per cow, however the grazing levels are adequate in order to maximize production per hectare.

-   In *Farm 6* this principle was completely adopted (mark=100%).

2) Marked Importance to Pasture Production:

-   *Farmer 6* stated that year-by-year, the productivity of the pastures is increasing; he thinks that they are utilizing 15 tonnes of dry matter of grass per hectare in their best paddocks. However, on average, pastures are producing approximately 8 tonnes of dry matter of grass.

-   *Farmer 6* and his son would like to apply more phosphate and nitrogen fertilizer to their pastures, however they have not been able to buy enough due to financial restrictions. The soils of the farm have a long history of cropping (onion and other crops) that was very extractive of nutrients. Another factor that affected the fertility of the soils was the levelling of the farm during the developing for flood-irrigation. Therefore *Farmer 6* thinks that in order to increase the fertility levels, the farm needs at least 1 tonne of the fertiliser called "phosphate bi-ammonic" (a fertilizer used in Argentina: 18 parts of Nitrogen, 46 parts of Phosphate and 0 parts of Potassium) per year. Last year they managed to buy and apply 500 kilograms of "phosphate bi-ammonic" per hectare (90 units of N and 230 units of P). These levels of fertilization are much higher than on typical Argentine farms.

-   Pasture is their main feed and, for *Farm 6,* increments in pasture production per hectare is the main driver of growth in milk production.

- In *Farm 6* this principle is adopted (mark=100%).

3) Quantitative Pasture Monitoring:

- The son of *Farmer 6*, who is in charge of the grazing management, does not monitor the pasture quantitatively. He manages the farm based on his experience; he estimates the pasture cover by mentally calculating how many days of grazing are still available. Therefore there is no historic recording of pasture cover and/or pasture growth in *Farm 6*.

- In *Farm 6* they do not use quantitative pasture monitoring probably because they believe they do not need to do it. The person that does the grazing management is the same person that monitors the pasture cover and growth, and does not think he needs to record it on paper or a computer spreadsheet. Pasture production is continually increasing therefore they do not give much value to records that would change the next year. Another reason is that they believe that doing quantitative pasture monitoring is not a priority for them and they prefer to spend their time on tasks that has a bigger impact on the farm results. Thus priorities may change when full pasture production is achieved.

- In *Farm 6* this principle was partially adopted (mark=50%).

4) Utilization of Formal Pasture Budgets:

- For the same or similar reasons that recording of pasture growth or cover was not adopted, and because the farm is relatively small (approximately 250 hectares) and the system is quite simple, the farmers believe that there is no need for complex calculations and budgets.

- In *Farm 6* this principle was not adopted (mark=0%).

5) Skilled and Motivated People Working on Farms:

- *Farmer 6* gained a Bachelor's degree in Agriculture and then did some postgraduate studies in New Zealand. *Farmer 6*'s son also gained a Bachelor's degree in Agriculture, and he has learnt from his father and from other farmers in New Zealand and Australia.

- The people that work on the farm (apart from *Farmer 6* and his son) are all employees; all of them receive wages for their work. Most of the employees have very low levels of formal education. However, they are moderately skilful because *Farmer 6* and his son work together with them in the daily farm activities and train them constantly.

- In *Farm 6* the milking of the cows takes place at 6 a.m. and at 5 p.m. They usually work many hours during the periods of calving and mating. Overall the working conditions are similar to those on New Zealand farms.

- In *Farm 6* the owners work their own farm, and their main income is from the farm. They are motivated because they are growing the family business. *Farmer 6* has become very skilful by learning from his own experience and from many successful examples in Argentina and several other countries. He has had the opportunity to learn what he calls *'the New Zealand dairy principles'*, and then test them in many different environments.

- *Farm 6* is probably the farm that was found to be more similar to a New Zealand farm in this aspect, especially because the owners work actively on the farm (mark=100%).

6)   Less than 15 cows per Set of Teat-cups, and Other Innovations that Impact on Labour Productivity:

- In *Farm 6* they have a policy of a maximum of 20 cows milked per set of teat-cups (1.33 times the number of cows per set of teat-cups of typical New Zealand farms). They try not to spend more than four hours per day milking the cows.

- There is one full-time person per approximately 50 cows, including the irrigation workers. However *Farmer 6* stated: *'this is a farm to be managed by 5 full-time persons with this number of cows* (actually there are 8 people working full-time and there are 400 cows), *but at the moment we are still developing the farm'*. When fully developed, they could milk approximately 750 cows (own calculations based on data provided by the farmer).

- *Farmer 6* mentioned that they invest in enough plant, machinery and buildings in order to work in a good environment, and in order to avoid spending too much time in duties that can be done more quickly with relatively inexpensive machinery. They are definitely trying to apply all New Zealand dairy principles if they can; however the socio-economic environment in which the farm is situated is very different from that in New Zealand. Some services and infrastructure that are common for New Zealand farmers (and also for some Argentine farmers in other regions) are not available for *Farm 6*, and consequently they need more people working on the farm.

- In *Farm 6*, this innovation was adopted to the highest degree of all case studies (mark=70%).

7)   Seasonal Calving, One or Two Calving Periods per Year:

- In *Farm 6* there are two calving periods, each of nine weeks, one in autumn and one in spring. They inseminate the cows for six weeks and then they put bulls in the herd for three more weeks.

- The family decided to have a seasonal calving system *'because the effort is concentrated.'* Farmer 6 mentioned that, for example, *'the mating is the most stressing period for my son and me, but it is no so stressful because we can delegate the monitoring of the calving herd and the other activities to the staff.'* However the heat detection and the insemination of the cows (during the mating) are not routine type activities and have to be done by *Farmer 6* or his son; they prefer to do these activities in a short period of time, concentrating their efforts.

- *'We have two calving periods in this farm because our client (the dairy company) needs winter milk. If it was only our decision we would have only one calving period in spring because it makes everything easier.'*

- They adopted this New Zealand innovation completely (mark=100%).

8)   New Zealand Genetics:

- In *Farm 6*, they started six years ago with average genetic merit Argentine Holstein heifers. Since then, they have been using only New Zealand semen; they used both Jersey and Holstein Friesian.

- The following quote expresses *Farmer 6's* thinking: *'I do not believe that it is necessary to have New Zealand genetics for having seasonal calving; however we are aware that it is easier with these genetics because they have been selected to calve each 12 months.'*

- *'We still have some cows of North American genetics, we bought 40 when we started with the dairy farm 6 years ago and 15 of them still are on the farm, the other 25 were culled because they were useless. But these 15 cows are very good producers and there have been seasons in which some of those cows were producing up to 24 litres at the time to dry them off, so we decided to keep milking them up to the next calving. For those cows probably you cannot define a season of 12 months, you need a different system.'*

- *'We rented some cows to some small farmers (5 to 15 cows) and all of them told me that they are good, that the New Zealand cows do not lose condition and they produce the same amount of milk. We also sold 75 heifers, many of them crossbreds, and the farmer who bought them is very happy with them, and they were the daughters of the lower producing cows.'*

- Therefore *Farm 6* adopted this New Zealand technology completely (mark=100%).

9) Rearing of Calves in Groups:

- In *Farm 6,* they have very good results rearing the calves *'we have 1 or 2 deaths per year* (out of approximately 350 calves reared per year), *and the moment of death is usually the calving'. Farmer 6,* who is in charge of rearing the calves, rears both male and female calves.

- *Farmer 6* mentioned that there are as many rearing systems as people in charge of rearing the calves. *'It all depends of the person in charge, some people rear very well with buckets and others rear very well with calf-teats and there are some that are a disaster with both.'*

- *'There are two main advantages of rearing with a barrel with calf-teats, one is that it takes less time and the other is that the calves do not grow like kids that did not go to the kindergarten, with this system the calves socialise very early. When you rear them individually, then put them into a group in a paddock, it takes them 2 or 3 weeks to realise that there are others like them and that they are part of a group, and this has an impact on their growth.'*

- *'The fact that it takes less total time to rear calves in groups is less important to me, what is more important is that this system gives the person more time to observe the calves, to see if they are sick and to touch their belly buttons to assess if they are infected. In the other system the person is so busy warming the milk, filling the buckets and moving the stakes that he or she does not have time to observe the calves.'*

- *'Another advantage is the psychology of the calf-teats that may have an impact. Also, the calves can move freely and go to shade if it is too hot, or shelter from the wind when it is too cold.'*

- They completely adopted this New Zealand practise for all the calves (mark=100%).

10) Style of Milking Shed and Milking System:

- In *Farm 6* the milking shed and milking system are New Zealand style (mark=100%).

Table 18: Summary of Innovations Adopted or Rejected by *Farm 6*

New Zealand Innovations	Mark
1) Focus on Production per Hectare	1
2) Marked Importance to Pasture Production	1
3) Quantitative Pasture Monitoring	0
4) Utilization of Formal Pasture Budgets	0
5) Skilled and Motivated People Working on Farms	1
6) Less than 15 cows per Set of Teat-cups	0.7
7) Seasonal Calving, One or Two Calving Periods per Year	1
8) New Zealand Genetics	1
9) Rearing of Calves in Groups	1
10) Style of Milking Shed and Milking System	1
TOTAL (out of 10)	**8.2**

7.1.7 Farm 7

General Characteristics

The Farm Business

Farm 7 is a husband and wife partnership. Both the husband and the wife direct the farm. The husband is more focused on the production activities of the farm and the wife does most of the paper work and financial management. Occasionally some of the children help with the farming activities. *Farm 7* is a dairy enterprise on leased land. The partnership owns all the cows, plant and the machinery.

The system is focused on maximizing milk production with the available resources. They have also been focused on increasing the cow numbers. *Farmer 7* mentioned that they usually have liquidity problems. This problem of cash availability forces them to use less concentrates and less fertilizer than they would like to use.

The Farmer

Farmer 7 is a very experienced person in farming. He worked for *Farmer 2* for several years and then they became equity partners in a dairy-farming project. *Farmer 7* bought half of the business from *Farmer 2* and continued with the dairy project that today is *Farm 7*.

Occasionally *Farmer 7* is involved in other dairy farms, especially during the development of new dairy farms, where he helps to supervise the daily duties and solve operational problems.

Farmer 7 is a local person and an active learner from colleagues, neighbours and friends. He very rarely reads technical journals.

Feeding Sources

On *Farm 7* farmers do not carry many physical records, therefore only general data about the system is detailed in this section.

Approximately 60% of the annual requirements of all the dairy animals are produced on the milk platform (grass, summer crops, winter crops and maize silage).

Figure 60: Feeds Eaten by *Farm 7* Dairy Animals

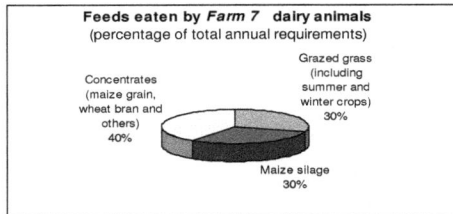

Feeds eaten by *Farm 7* dairy animals
(percentage of total annual requirements)

Concentrates (maize grain, wheat bran and others) 40%

Grazed grass (including summer and winter crops) 30%

Maize silage 30%

New Zealand Innovations Adopted and/or Rejected

1) Focus on Production per Hectare:

- *Farmer 7* was clear that he tries to keep as many cows as possible because he has the dream of starting another dairy farm business in the future.

- The main fixed cost is the cost of leasing the farm; therefore *Farmer 7* tries to produce as much milk per hectare as possible in order to reduce the cost of leasing per kilogram of milk produced.

- On *Farm 7* this principle was completely adopted (mark=100%).

2) Marked Importance to Pasture Production:

- On *Farm 7,* they do not fertilize their pastures. They use some fertilizer only at the time of planting them.

- *Farmer 7* gives priority to cows over pastures, and they usually graze very low in order to keep the pastures even (without patches of longer grass). Only in spring do they leave slightly longer residuals.

- Even though *Farmer 7* believes that grass is the cheapest feed, he prefers to invest in buying supplements (for example wheat bran, which is relatively inexpensive in Argentina) or making maize silage, rather than fertilizing the pastures. This is because he finds that the response of pastures to fertilization is more uncertain than buying or making supplements.

- Another important aspect of *Farm 7* is that there are usually feed deficits in the winter when pastures do not grow, even if they are fertilized; consequently they prefer to use maize silage and supplements instead.

- On *Farm 6* this principle was not adopted (mark=0%).

3)   Quantitative Pasture Monitoring:

- *Farmer 7* and the person in charge of the grazing are very experienced in grazing management. They do not use any kind of recording method for pasture growth or pasture cover.

- In *Farm 7* they do not use quantitative pasture monitoring because they believe they that they do not need to do it.

- In *Farm 7* this principle was not adopted (mark=0%).

4)   Utilization of Formal Pasture Budgets:

- *Farmer 7* believes that he and the person in charge do not need to do pastoral or feed budgets. They base their feed management on the observation of the farm and on their own experience.

- In *Farm 7* this principle was not adopted (mark=0%).

5)   Skilled and Motivated People Working on Farms:

- *Farmer 7* finished primary and secondary education; the secondary school he attended included a specialization in farming. He then worked for many years in farming and accumulated experience, knowledge and skills. Most of his experience has been in operational and tactical planning, implementation and control. When they started with *Farm 7*, he and his wife became involved in the strategic aspects of the farm business. When *Farmer 7* is not employed as a manager in other dairy farms he works on *Farm 7*.

- The other people that work on *Farm 7* have quite low levels of formal education. Two of them are very experienced in dairy farming and have worked with *Farmer 7* for many years.

167

- In *Farm 7* the owners work in their own farm business, and their main income comes from the farm. They are strongly motivated because they are growing the family business. *Farmer 7* has experienced the adoption of the New Zealand innovations described in *Farm 2*, and has adopted the practices that he believes are beneficial for his own farm.

- The working environment and the working conditions are better than on a typical Argentine dairy farm. *Farmer 7* mentioned that one of his daughters works occasionally on the farm and is planning to work full time next season.

- In this aspect *Farm 7* is similar to a New Zealand dairy farm (mark=100%).

6) Less than 15 cows per Set of Teat-cups, and Other Innovations that Impact on Labour Productivity:

- On *Farm 7* they have approximately 25 cows per sets of teats-cups (1.67 times more cows per set of teat-cups than in New Zealand). *Farmer 7* does not like situations when people have to spend too much time milking the cows.

- In *Farm 7* they usually work very intensively and sometimes long hours. The fact that the owner (the person who makes the investment decisions) works on the farm has helped to design a simple system in which everything is done as quickly as possible (mark=50%).

7) Seasonal Calving, One or Two Calving Periods per Year:

- On *Farm 7* there are two calvings, each for a period of three months, one in autumn (70% of the cows) and one in spring (30% of the cows). They started with the cows that they had, and slowly they have fitted the cows into the two seasons.

- *'We need the cows to produce during summer (spring calving) because we still have to pay the farm lease and other costs. And our cows produce during the winter (autumn calving) because the dairy company indirectly obliges us to do so because the milk price is higher in winter.'*

- They adopted this New Zealand innovation completely (mark=100%).

8)   New Zealand Genetics:

-    In *Farm 7*, they are slowly increasing their utilization of New Zealand genetics. They bought some Jersey bulls from *Farmer 4* and last year approximately 30% of the cows were mated with New Zealand genetics (both bulls and semen). *'My idea is to keep increasing the proportion of New Zealand genetics, my idea is not to have all Jersey cows but I would like to have crossbreds or three fourths Jerseys.'*

-    *Farmer 7* mentioned that he likes New Zealand genetics cows because of *'their size, the strength of their legs and their udders'*. He mentioned that after working for many years with New Zealand genetics, he started working in a dairy farm with all North American Holstein cows. He said that he tried to send the cows to graze in paddocks that were quite far from the milking shed and the cows started to *'be destroyed'* and that *'in summer they stayed under the trees all day while my crossbreds at home were grazing comfortably.'* I suggested to the owners of this farm that they should buy some New Zealand Holstein Friesian cows, they did it, and they had good results with them: *'the New Zealand Holstein Friesian cows did not produce as much as the North American cows but they got in-calf every year (and the others did not) and produced similar amounts of milksolids.'*

-    *Farm 7* adopted this New Zealand technology partially (mark=30%).

9)   Rearing of Calves in Groups:

-    *Farmer 7* believes that rearing calves in groups is much better than doing it individually. He mentioned that in a farm in which they were having problems rearing the calves, he suggested changing to group-rearing and all problems were solved. *Farmer 7* was asked if rearing calves individually would be successful in situations in which the person in charge of rearing the calves was not very skilful. He answered that people can be easily trained to rear calves in groups, and also stated that he does not believe in having people that are not skilful and do not want to learn. *'If that is the situation the most that you can do is wait one or two weeks until you find another worker'*.

- *In Farm 7* they completely adopted this New Zealand practise for all the calves (mark=100%).

10)   Style of Milking Shed and Milking System:

- On *Farm 7* they have a typical New Zealand herringbone with a one high-line of sets of teat cups because they believe is a simpler and less expensive system.

- Because *Farm 7* is on leased land; *Farmer 7* prefers not to invest too much in buildings, therefore they used a shed that was already on the farm. When the leasing contract finishes they would take most of the investment with them. In *Farm 7* the milking shed and milking system are New Zealand style (mark=100%).

Table 19: Summary of Innovations Adopted or Rejected by *Farm 7*

New Zealand Innovations	Mark
1) Focus on Production per Hectare	1
2) Marked Importance to Pasture Production	0
3) Quantitative Pasture Monitoring	0
4) Utilization of Formal Pasture Budgets	0
5) Skilled and Motivated People Working on Farms	1
6) Less than 15 cows per Set of Teat-cups	0.5
7) Seasonal Calving, One or Two Calving Periods per Year	1
8) New Zealand Genetics	0.3
9) Rearing of Calves in Groups	1
10) Style of Milking Shed and Milking System	1
TOTAL (out of 10)	**5.8**

7.2 Summary of Adoption of NZ Innovations by the Case Study Farms

Figure 61: Summary of Adoption of NZ Innovations by the Case Study Farms

	Farm 1	Farm 2	Farm 3	Farm 4	Farm 5	Farm 6	Farm 7	
1) Focus on Production per Hectare	1	1	1	1	1	1	1	**7.0**
2) Marked Importance to Pasture Production	0.25	1	0.6	1	0	1	0	**3.9**
3) Quantitative Pasture Monitoring	0.25	0.9	0.8	0.9	0	0.5	0	**3.4**
4) Utilization of Formal Pasture Budgets	0	1	0.9	1	0	0	0	**2.90**
5) Skilled and Motivated People Working on Farms	0	1	0	0.8	0.3	1	1	**4.1**
6) Less than 15 cows per Set of Teat-cups	0.45	0.5	0.3	0.4	0.0	0.7	0.5	**2.85**
7) Seasonal Calving, One or Two Calving Periods per year	0.17	1	0	1	1	1	1	**5.2**
8) New Zealand Genetics	0.2	1	0	1	0.67	1	0.3	**4.2**
9) Rearing of Calves in Groups	0.1	1	1	1	1	1	1	**6.1**
10) Style of Milking Shed and Milking System	0.8	1	1	1	1	1	1	**6.8**
	3.2	**9.4**	**5.6**	**9.1**	**5.0**	**8.2**	**5.8**	

Figure 61 shows a summary of the innovations adopted by the case study farms. The bold numbers of the bottom of the figure are the sum of all the innovations adopted by each of the farms. For example *Farm 2* with 9.4 innovations adopted (out of 10) is the case study that adopted more New Zealand innovations. The bold numbers at the extreme right of the figure are the sum of the different proportions of the same innovation adopted by the case studies. For example the *Focus on Production per Hectare* (innovation 1) was adopted by 7 (out of 7) of the case studies. In the same way *Farm 1* is the case study that less innovations adopted, and innovations 4 (*Utilization of Formal Pasture Budgets*) and 6 (*Less than 15 cows per Set of Teat-cups*) were the innovations less adopted by the case studies.

7.3   The Context: Perception of the Attributes of the Innovations

In order to have a general idea of the perception of some New Zealand innovations by Argentine dairy farmers and consultants, an informal survey was done. The survey asked about the perception of the interviewee of a group of defined New Zealand innovations. Each innovation was ranked for 4 characteristics, which were: "Level of Advantage", "Level of Compatibility", "Level of Trialabilty", and "Level of Awareness of Results" (these concepts are explained further on section 2.1.3.1). The ranking was from "1" (very low) to "4" (very high).

Is important to mention that the questionnaires included 14 New Zealand innovations, these were the innovations that were proposed in the beginning of the present study and were then reduced to the final 10 innovations (see section 6.1).

In total 29 questionnaires were answered by three different groups of Argentine people:

1) The 7 case study farmers and some of their dairy farmers answered 14 questionnaires.

2) Four Argentine dairy farmers with no known relationship with New Zealand innovations were surveyed.

3) And finally 9 consulting officers were surveyed.

The average results of the perception survey are summarized in the following tables for each group:

Table 20: Results from the Perception Survey with Case Study

Farmers (n=14)

	Level of Advantage	Level of Compatibility	Level of Trialability	Level of Observability
1) Stocking rate decided taking into account prices of milk and supplements	3	3	3	3
2) Intention to raise the phosphate level of soils	3	4	4	3
3) Intention to have pastures that last more than 4 years without the need to replant them	4	3	3	3
4) Utilization of adequate species on pastures	3	3	3	3
5) Quantitative pasture monitoring	4	4	3	4
6) Target pre and post grazing pasture covers	3	3	3	3
7) Utilization of formal pastoral budgets	4	4	3	4
8) More "technology" and less working hours	4	4	4	3
9) Skilled and motivated people working on farms	4	4	3	4
10) Working contract similar to New Zealand share-milking contracts	3	3	3	4
11) Seasonal calving, one or two calving periods	4	3	3	3
12) New Zealand genetics	4	4	3	3
13) Rearing of calves in groups	4	4	4	4
14) Style of milking shed and milking system	4	3	3	3
TOTAL	49	48	44	48

GENERAL TOTAL = 189

Table 21: Results from the Perception Survey with Argentine Dairy
Farmers (n=4)

	Level of Advantage	Level of Compatibility	Level of Trialability	Level of Observability
1) Stocking rate decided taking into account prices of milk and supplements	3	3	3	3
2) Intention to raise the phosphate level of soils	3	4	3	4
3) Intention to have pastures that last more than 4 years without the need to replant them	4	3	3	4
4) Utilization of adequate species on pastures	3	3	3	3
5) Quantitative pasture monitoring	4	4	3	4
6) Target pre and post grazing pasture covers	4	4	4	4
7) Utilization of formal pastoral budgets	4	4	4	4
8) More "technology" and less working hours	3	3	3	3
9) Skilled and motivated people working on farms	3	3	3	3
10) Working contract similar to New Zealand share-milking contracts	2	2	2	2
11) Seasonal calving, one or two calving periods	2	3	3	3
12) New Zealand genetics	2	3	3	3
13) Rearing of calves in groups	2	3	4	4
14) Style of milking shed and milking system	2	3	3	2
TOTAL	41	42	41	42

GENERAL TOTAL = 166

Table 22: Results from the Perception Survey with Argentine Dairy Consultants (n=9)

	Level of Advantage	Level of Compatibility	Level of Trialability	Level of Observability
1) Stocking rate decided taking into account prices of milk and supplements	3	3	3	3
2) Intention to raise the phosphate level of soils	4	4	3	4
3) Intention to have pastures that last more than 4 years without the need to replant them	3	3	3	3
4) Utilization of adequate species on pastures	3	3	3	3
5) Quantitative pasture monitoring	4	3	4	3
6) Target pre and post grazing pasture covers	3	3	3	3
7) Utilization of formal pastoral budgets	4	4	4	4
8) More "technology" and less working hours	3	3	3	3
9) Skilled and motivated people working on farms	4	4	3	4
10) Working contract similar to New Zealand share-milking contracts	3	3	2	3
11) Seasonal calving, one or two calving periods	3	2	3	3
12) New Zealand genetics	2	2	2	2
13) Rearing of calves in groups	3	3	3	3
14) Style of milking shed and milking system	3	3	3	3
TOTAL	45	43	42	43

GENERAL TOTAL = 173

8 RESULTS (Part 2)

8.1 Performance of The Case Studies: Association with Adoption of New Zealand Innovations

In this section some physical and financial indicators of the case study farms are analysed and contrasted with typical New Zealand and Argentine farms.

8.1.1 Definition of the seasons to be analysed

Data from two seasons is presented for each farm. It is important to mention that the seasons are considered in different ways by the farms (see Table 23).

The differences in the financial years meant data collected from Argentine farms varied and this was especially the case with milk prices. The variation was exacerbated in 2002 by high inflation and by a significant devaluation of the Argentine currency. Therefore when comparing the Argentine farms, the differences in financial years should be remembered.

Table 23: Definition of the Financial Years of the Case Studies

	Season 2002	Season 2003
AR-150 (IFCN model)	January 02 – December 02	January 03 – December 03
AR-350 (IFCN model)	January 02 – December 02	January 03 – December 03
AR-1400 (IFCN model)	January 02 – December 02	January 03 – December 03
Farm 1	July 01 – June 02	July 02 – June 03
Farm 2	February 02 – January 03	February 03 – January 04
Farm 3	July 01 – June 02	July 02 – June 03
Farm 4	July 01 – June 02	July 02 – June 03
Farm 5	February 02 – January 03	Not enough data available
Farm 6	Not enough data available	Not enough data available
Farm 7	Not enough data available	July 02 – June 03
NZ-239 (IFCN model)	June 01 – May 02	June 02 – May 03
NZ-447 (IFCN model)	June 01 – May 02	June 02 – May 03
NZ-835 (IFCN model)	June 01 – May 02	June 02 – May 03

Based on the financial years used by the Argentine farms, there are three distinctive groups for each season:

Season 2002

The first group is formed by the three typical Argentine farms with original data from January 2002 to December 2002. During this period the exchange rate changed from 1 AR-$ per each US-$ at the beginning of January, to 3.5 AR-$ per each US-$ at the end of December, with an average exchange rate of 3.1 AR-$ per each US-$ (*FXHistory: Historical currency exchange rates*, 2005). The inflation during this period was 118% (calculations based on the Wholesale Price Index INDEC, 2005). Argentine milk prices during this period increased 2.5 times in AR-$ (calculations based on Argentine Milk Prices).

The second group is formed by three case studies with original data from July 2001 to June 2002; they are AR-6150, AR-1754 and 1483 (*Farms 1, 3* and *4,* respectively). These farms share six months in common with the farms of the first group. The exchange rates during this period changed from 1 AR-$ per each US-$ in July 2001 to 3.63 AR-$ per each US-$ in June 2002, with an average exchange rate of 1.8 AR-$ per each US-$ (*FXHistory: Historical currency exchange rates*, 2005). Inflation during this period was 88% (calculations based on the Wholesale Price Index INDEC, 2005). Argentine milk prices increased during this period 1.4 times in AR-$ (calculations based on Argentine Milk Prices).

The third group is formed by AR-2087 and AR-915 (*Farms 2* and *5,* respectively). These two farms have their financial year from February 2002 to January 2003. And they differ only by one month from the financial year of the first group and share five months in common with the financial year of the second group. The exchange rates during this period changed from 2 AR-$ in February 2002 per each US-$ to 3.3 AR-$ per each US-$ in January 2003, with an average exchange rate of 3.3 AR-$ per each US-$ (*FXHistory: Historical currency exchange rates*, 2005). Inflation during this period was 106% (calculations based on the Wholesale Price Index INDEC, 2005). Argentine milk prices increased during this period 2.5 times in AR-$ (calculations based on Argentine Milk Prices).

The farms of the three different groups experienced quite distinct changes in exchange rates, inflation, and Argentine milk prices. Therefore, for the 2002 season, financial comparison across groups of farms were avoided and comparisons within groups were done cautiously.

Season 2003

First group:

- Formed by the three typical Argentine farms.

- Original data from January 2003 to December 2003.

- Exchange rate with little variation and average of 3 AR-$ per each US-$ (*FXHistory: Historical currency exchange rates*, 2005).

- Inflation for the period 2% (calculations based on the Wholesale Price Index INDEC, 2005).

- Argentine milk prices relatively constant, with seasonal variations, in AR-$ (calculations based on Argentine Milk Prices).

Second group:

- Formed by AR-6350, AR-1700 and AR-1483 (*Farms 1, 3* and *4*, respectively). AR-400 (*Farm 7*) is added to this group in this season.

- Original data from July 2002 to June 2003.

- Exchange rate of 3.6 AR-$ per each US-$ in July 2002 to 3 AR-$ per each US-$ in June 2003, with an average of 3.3 AR-$ per each US-$ (*FXHistory: Historical currency exchange rates*, 2005).

- Inflation for the period 8.1% (calculations based on the Wholesale Price Index INDEC, 2005).

- Argentine milk prices increased 1.92 times in AR-$. Milk prices for the first group were, on average, 1.29 times higher (in US-$) than for this group (calculations based on Argentine Milk Prices).

Third group:

- Only AR-2530 (*Farm 2*) forms this group in this season.

- Original data from February 2003 to January 2004.

- Exchange rate with little variation and average of 3 AR-$ per each US-$ (*FXHistory: Historical currency exchange rates*, 2005).

- Inflation 1.2% (calculations based on the Wholesale Price Index INDEC, 2005).

- This group had a period of higher Argentine milk prices than the other two groups; milk prices were 1.02 times higher (in US-$) than for the first group and 1.32 times higher (in US-$) than the for second group (calculations based on Argentine Milk Prices).

Difference across the financial years is lower in this season than in 2002, especially for the first and third groups, which experienced less variation in exchange rates and inflation. Therefore, comparisons across the first and AR-2530 or *Farm 2* (third group) were done, taking into account the small difference in the Argentine milk price between the two financial years. And comparisons of farms of the second group with farms of the other two groups was made with caution, taking into account the difference in the Argentine milk price between the two financial years.

8.1.2 The Case Studies in IFCN Format

Basic descriptive data for the case study farms in 2003 are shown in Table 10 (Chapter 5). A similar format to that used in section 3.2 to present the results from comparisons between Argentine and New Zealand typical farm models is used in this chapter to present the results of analyses for the case study farms.

As shown in Table 19, *Farm 6* is not presented because there were not enough data for the analysis.

Following the IFCN format, farms are named with two letters of the country and the average number of cows for that season (see Table 24).

Table 24: Names Given to the Case Study Farms on Figures

	Season 2002	**Season 2003**
Farm 1	AR-6150	AR-6350
Farm 2	AR-2087	AR-2530
Farm 3	AR-1754	AR-1700
Farm 4	AR-1375	AR-1483
Farm 5	AR-915	-
Farm 7	-	AR-400

As shown in Table 24, *Farm 5* (AR-915) appears only in the analysis for the 2002 season, and *Farm 7* (AR-400) appears only in the analysis for the 2003 season.

8.1.3 Financial Performance

Milk Prices, Profits, Operating Profit Margin, and Returns on Investment (ROI) are analysed in order to compare the financial performance of the Argentine farms. First the indicators are analysed for 2002, and then 2003. Finally, the level of financial performance is related to the level of adoption of New Zealand innovations.

Season 2002

Figure 62: Milk Price (season 2002)

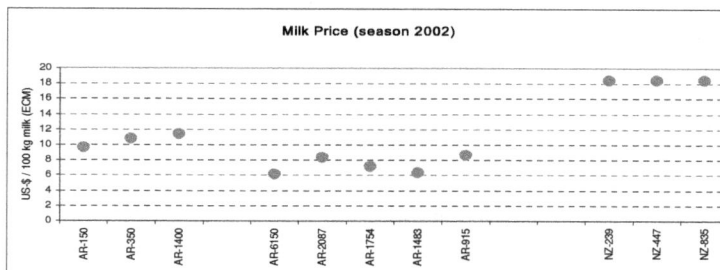

Milk prices are calculated, in the IFCN database, by dividing the milk returns by the total kilograms of milk (ECM) sold.

The three Argentine typical farms (first group) had similar milk prices, slightly increasing with the farms' size.

Farms AR-6150, AR-1754 and AR-1483 (*Farms 1, 3* and *4* respectively; second group) all had similar milk prices, which were significantly lower than the milk prices of the first group.

Farms AR-2087 and AR-915 (*Farms 2* and *5* respectively; third group) all had very similar milk prices, which were lower than milk prices of the first group and higher than the second group.

Differences in milk prices across the three groups were not due to differences in the production systems, milk quality, or negotiating power; they were caused mainly by the different effects of the variations of the exchange rates, inflation, and milk prices for the different financial years operated by the three groups of farms.

However, differences between farms of the same groups are related to negotiating power, quantity of milk produced, quality of milk, milk composition, dairy company to which the farm is selling, proportion of milk sold per month[24], and other factors.

Figure 63: Returns and Profits (season 2002)

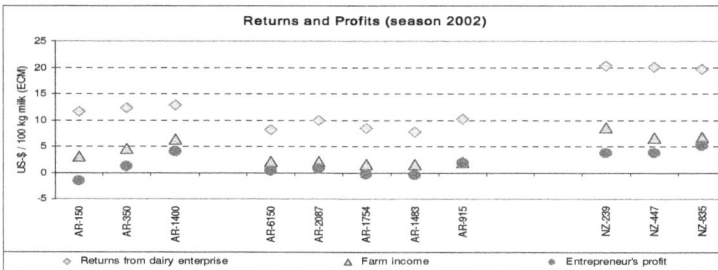

See section 4.3 for the definition of: Returns from the dairy enterprise, Farm Income and Entrepreneur's Profit.

As was previously mentioned, the Entrepreneur's Profit measures the economic sustainability of the business in the long run (IFCN, 2002). Figure 64 shows that AR-1400 is the farm with higher economic sustainability of the first group. Within the second group (Farms AR-6150, AR-1754 and AR-1483) only AR-6150 (*Farm 1*) had a

---

[24] The fact that some farms are seasonal and produce a bigger proportion of milk during the months in which the milk price is higher can result in a higher average annual price.

positive Entrepreneurs' Profit. And within the third group AR-915 (*Farm 5*) had higher Entrepreneurs Profit than AR-2087 (*Farm 2*).

Figure 64: Operating Profit Margin (season 2002)

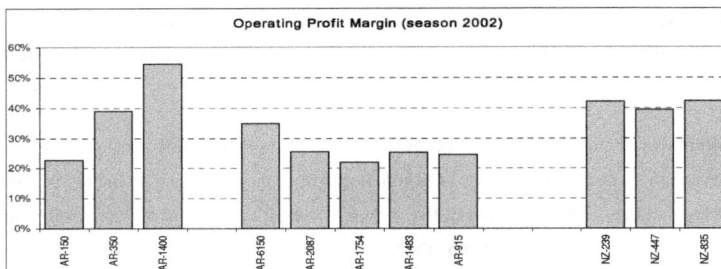

The Operating Profit Margin is calculated by dividing the Operating Profit (calculated as Farm Income + Paid land rent + Paid Interest on liabilities – Calculated cost for unpaid family labour, see section 4.3) by the Total Returns. As was previously mentioned, the Operating Profit Margin is an indicator of the operating efficiency of the dairy farms; it indicates how well the farms have turned income into profit.

Within the first group (the three typical Argentine farms), the Operating Profit Margin increased with the size of the farms. Within the second group (AR-6150, AR-1754 and AR-1483), AR-6150 (*Farm 1*) had the highest Operating Profit Margin. Within the third group (AR-2087 and AR-915), AR-2087 had the highest Operating Profit Margin.

Figure 65: Return on Investment (season 2002)

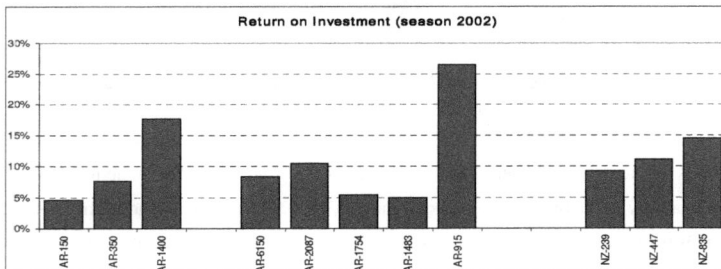

The Return on Investment (ROI) is calculated by dividing the Operating Profit by the market value of all the assets invested in the business. It is a relevant indicator for people who want to invest in dairy farming.

Within the first group (the three typical Argentine farms), in season 2002, AR-1400 had the highest ROI. Within the second group of farms (Farms AR-6150, AR-1754 and AR-1483), AR-6150 (*Farm 1*) had the highest ROI. And within the third group AR-915 (*Farm 5*) had much higher ROI than AR-2087 (*Farm 2*); this is caused by the fact that *Farm 5* leased all the land and consequently had less total assets invested.

Season 2003

Figure 66: Milk Price (season 2003)

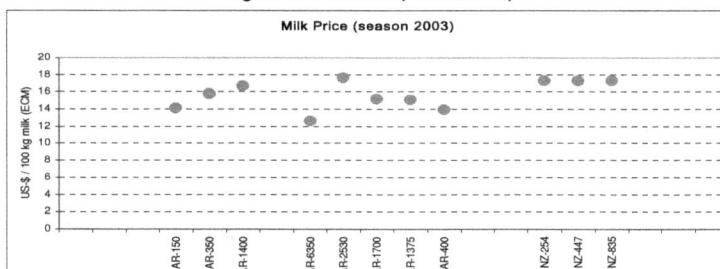

For the 2003 season the average milk price for the typical Argentine farms (first group) was 15.5 US-$ per 100 kilograms of milk (ECM). For the second group the average milk price was 14.7 US-$ per 100 kilograms of milk (ECM) for AR-1700, AR-1375 and AR-400 (*Farms 3, 4* and *7*) and AR-6350 (*Farm 1*) had a lower milk price (12.6 US-$ per 100 kilograms of milk). AR-2530 (*Farm 2*), which is the only farm of group three, had a milk price of 16.7 US-$ per 100 kilograms of milk.

As was previously mentioned differences in milk prices between farms of different groups are mainly due to the distinctive financial years and differences between farms of the same groups are related other factors.

Within the second group AR-6350 (*Farm1*), had a considerably lower price than the average of the other three case studies. The causes are unknown but this could be related to a difference in the way of recording the financial information, because *Farm 1* also presented lower expenses than the other farms.

When analysing other financial indicators the following points should be remembered:

- AR-2530 had the highest milk price received in this season (16.7 US-$ per 100 kilograms of milk).

- AR-2530 milk price was 1.25, 1.12 and 1.06 times higher than AR-150, AR-350 and AR-1400, respectively (first group).

- AR-2530 milk price was 1.40, 1.16, 1.17 and 1.27 times higher than AR-6350, AR 1700, AR-1375 and AR-400, respectively (second group).

Figure 67: Returns and Profits (season 2003)

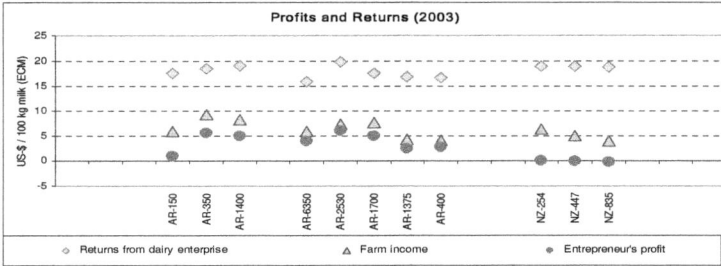

AR-350 had the highest Entrepreneurs' Profit within the first group of typical Argentine farms. AR-1700 (*Farm 3*) had the highest Entrepreneurs' Profit within the second group (AR-6350 or *Farm 1*, AR-1375 or *Farm 5* and AR-400 or *Farm 7*). AR-2530 or *Farm 2* (third group) had an Entrepreneurs' Profit 1.07 times of that of AR-350.

Figure 68: Operating Profit Margin (2003)

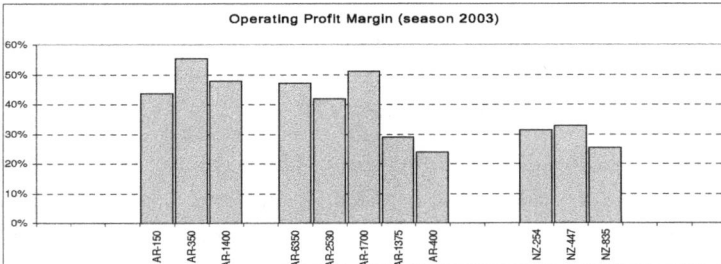

Within the first group (the three typical Argentine farms), AR-350 had the highest Operating Profit Margin. Within the second group (AR-6350, AR-1700, AR-1375 and AR-400), AR-1700 (*Farm 3*) had the highest Operating Profit Margin. AR-350 had an Operating Profit Margin 1.32 times higher than AR-2530 or *Farm 2* (third group).

Figure 69: Return on Investment (season 2003)

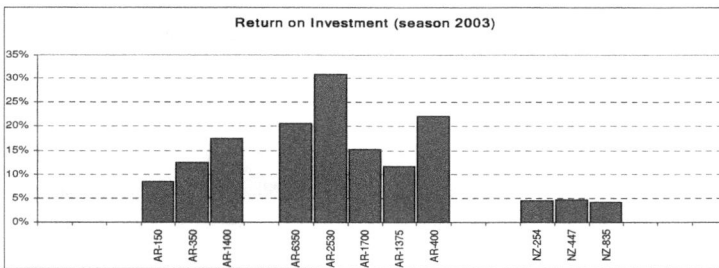

Within the first group (the three typical Argentine farms), AR-1400 had the highest ROI. Within the second group (AR-6350, AR-1700, AR-1375 and AR-400), AR-400 (*Farm 7*) had the highest ROI; this is caused by the fact that AR-400 leased all the land, and therefore, had less investment in the business. The second best ROI within the second group was to AR-6350 (*Farm 1*). AR-2530 or *Farm 2* (third group) had an ROI 1.5 and 1.77 times higher than AR-1400 and AR-6350 (*Farm 1*), respectively.

Summary for Season 2002:

Within the first group, AR-1400 had the highest Entrepreneurs' Profit, Operating Profit Margin, and Return on Investment.

Within the second group, AR-6150 (*Farm 1*) had the highest Entrepreneurs' Profit, Operating Profit Margin, and Return on Investment.

Within the third group, AR-915 (*Farm 5*) had the highest Entrepreneurs' Profit; and AR-2087 (*Farm 2*) had the highest Operating Profit Margin. The ROI of AR-915 (*Farm 5*) was much higher than that of AR-2087 (*Farm 2*), due partly to the fact that AR-915 leased all the land.

Summary for Season 2003:

Within the first group, AR-350 had the highest Entrepreneurs' Profit and Operating Profit Margin, but AR-1400 had the highest ROI.

Within the second group, AR-1700 (*Farm 3*) had the highest Entrepreneurs' Profit and Operating Profit Margin. And AR-6350 (*Farm 1*) had the highest ROI of the farms on owned land. AR-400 (*Farm7*) had the highest ROI of all the farms of the second group due to the fact that it leased all the land.

AR-2530 (*Farm 2*) had a higher Entrepreneurs' Profit (1.07 times) than AR-350, however this could be influenced by the fact that AR-2530 received a milk price that is 1.12 times higher than the milk price received by AR-350. AR-2530 (*Farm 2*) had the highest ROI, 1.5 and 1.77 times higher than AR-1400 and AR-6350 (*Farm 1*), respectively. However it should be remembered that AR-2530 received a milk price 1.06 and 1.16 times higher than AR-1400 and AR-6350 respectively.

General Analysis of Financial Performances:

For the first group, AR-1400 performed better in 2002. And in 2003, AR-350 and AR-1400 performed similarly.

For the second group, AR-6150 (*Farm 1*) had the best performance in 2002. And in 2003, AR-1700 (*Farm 3*) and AR-6350 (*Farm 1*) performed similarly.

For the third group, AR-915 (*Farm 5*) and AR-2087 (*Farm 2*) performed similarly. And in 2003, AR-2530 (*Farm 2*) performed better than any farm in Entrepreneurs' Profit and ROI. The better performance of AR-2530 was helped by the fact that it had a financial year in which the Argentine milk prices were higher.

Adoption of New Zealand Innovations and Financial Performance:

The number of New Zealand innovations adopted by the farms analysed in this section are shown in the following table:

Table 25: Number of New Zealand Innovation Adopted (out of 10)

AR-150	AR-350	AR-1400	Farm 1	Farm 2	Farm 3	Farm 4	Farm 5	Farm 7
0	0	0	3.2	9.4	5.6	9.1	5.0	5.8

The following figures show the association between the level adoption of New Zealand innovations and three financial indicators (Entrepreneurs' Profit, Operating profit Margin, and ROI). On the X-axis the number of innovations adopted are shown, and in the Y-axis the values for the financial indicators are shown.

Figure 70: Number of New Zealand Innovations Adopted and Entrepreneurs' Profit (season 2003)

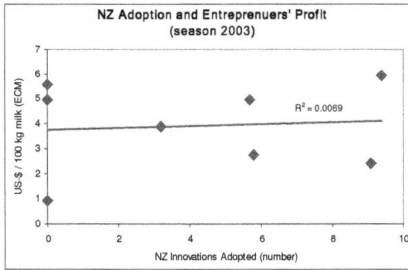

No association was found between the number of innovations adopted by the Argentine farms and their Entrepreneur's Profit in 2003. *Farm 2* (AR-2087 in 2002, and AR-2530 in 2003) was one of the best performers and was the farm that adopted the highest number of innovations (see Table 25). However AR-1400, AR-350 and *Farm 1* (AR-6150 and AR-6350), had similar financial performances and adopted very few New Zealand innovations (see Table 25).

**Figure 71: Number of New Zealand Innovations Adopted and
Operating Profit Margin (season 2003)**

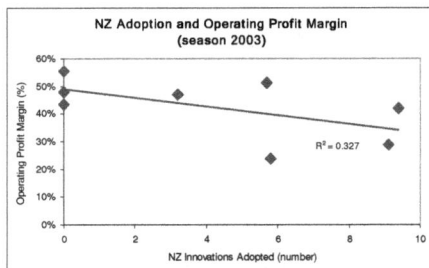

An association was found between the number of innovations adopted by the Argentine farms and their Operating Profit Margin in 2003. Figure 71 shows that, in general, the farms that adopted more New Zealand innovations had lower Operating Profit Margin in 2003.

**Figure 72: Number of New Zealand Innovations Adopted and
Return on Investment (season 2003)**

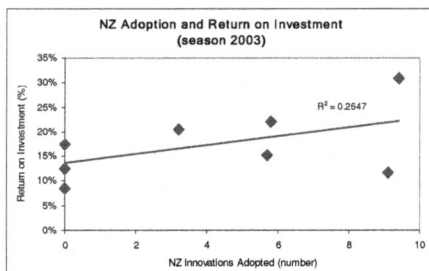

An association was found between the number of innovations adopted by the Argentine farms and their Return on Investment in 2003. Figure 72 shows that, in general; the farms that adopted more New Zealand innovations had higher Return on Investment in 2003.

8.1.4 Total Costs

Figure 73: Total Costs and Returns of the Dairy Enterprise (season 2002)

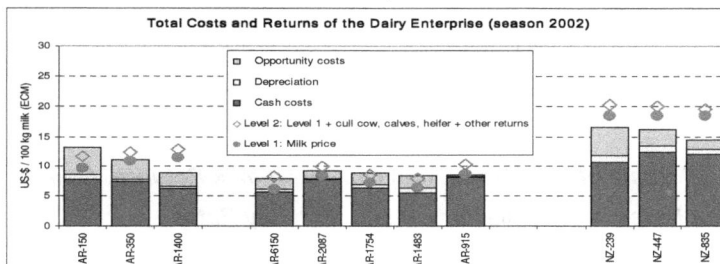

In 2002 (see Figure 73), within the farms of the first group (the three typical Argentine farms) AR-1400 had the lowest Total Costs per 100 kilograms of milk (ECM) produced. Within the second group (Farms AR-6150, AR-1754 and AR-1483), AR-6150 (*Farm 1*) had the lowest Total Costs but this was only slightly lower than for the two other farms. Within the third group AR-915 (*Farm 5*) had lower Total Costs than AR-2087 (*Farm 2*).

The difference in Total Costs across the two groups of case studies is lower than the difference in Milk Prices across the same two groups. As was previously mentioned the Total Costs were less affected by the financial distortions than the Milk Prices. In general Total Costs per 100 kilograms of milk (ECM) in 2002 were lower for the case studies than for the typical farms.

Figure 74: Total Costs and Returns of the Dairy Enterprise (season 2003)

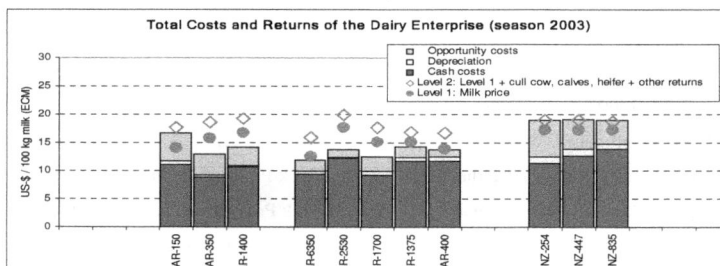

In 2003 (see Figure 74), within the farms of the first group (the three typical Argentine farms) AR-350 had the lowest Total Costs per 100 kilograms of milk (ECM) produced. Within the second group (Farms AR-6350, AR-1700, AR-1375 and AR-400), AR-6350 (*Farm 1*) had the lowest Total Costs. AR-2530 or *Farm 2* (third group) had lower Total Costs than the typical farms (first group), and was very similar to AR-400 (*Farm 7*) of the second group.

8.1.5 Costs of Milk Production Only, Non-milk Returns, Animal Purchases, and Replacement and Mortality Rates

For all Argentine farms, and for both seasons, the Costs of Milk Production Only per 100 kilograms of milk were lower in Argentina than in New Zealand. As was previously explained this started to happen after the devaluation of the Argentine currency in January 2002.

Season 2002

Figure 75: Costs of Milk Production Only (season 2002)

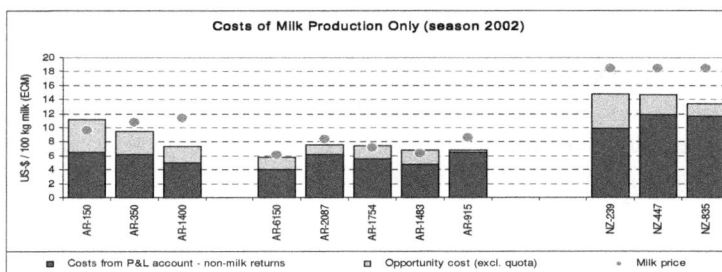

In 2002 (see Figure 75), within the farms of the first group (the three typical Argentine farms), AR-1400 had the lowest Costs of Milk Production Only per 100 kilograms of milk (ECM) produced. Within the second group (Farms AR-6150, AR-1754 and AR-1483), AR-6150 (*Farm 1*) had the lowest Costs of Milk Production Only. Within the third group AR-915 (*Farm 5*), had lower Costs of Milk Production Only than AR-2087 (*Farm 2*); however *Farm 2* had lower 'Costs from Profit & Loss Account – Non-milk

Returns' than *Farm 5*. The fact that *Farm 5* had less Opportunity Costs is because this case study farm leased all its land.

Figure 76: Non-milk Returns (season 2002)

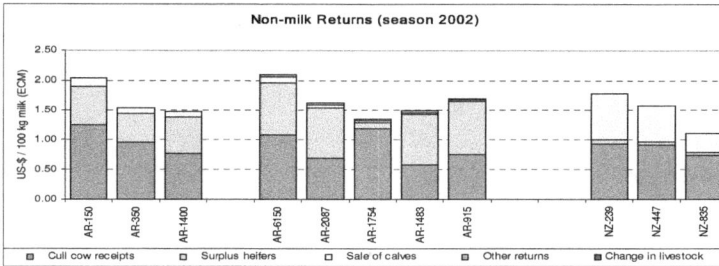

Figure 77: Animal Purchases (season 2002)

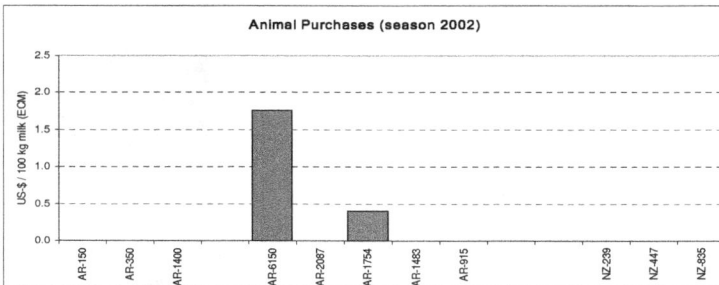

Figures 76 and 77 have to be analysed together because the Animal Purchases are not deducted from Non-milk Returns, although this is relevant only for AR-6150 and AR-1754.

Within the farms of the first group (the three typical Argentine farms) AR-150 had the highest Non-milk Returns per 100 kilograms of milk (ECM) produced. This is possibly due to the fact that the other two typical farms have, on average, 1.4 times more milk production per cow.

Within the second group (Farms AR-6150, AR-1754 and AR-1483), AR-6150 (*Farm 1*) had the highest Non-milk Returns; however if the Animal Purchases are deducted then it had the lowest Non-milk Returns. *Farm 1* (AR-6150) had so high Animal Purchases because it has a policy of selling all the heifers calves when they are around 6 months old to a 'Heifer Raising Enterprise' and buy all the replacement cows from this enterprise. Therefore *Farm 5* had the highest Non-milk Returns of this group, this is possibly due to lower mortality rates of calves, heifers and cows.

Within the third group AR-915 (*Farm 5*) had lower Costs of Milk Production Only than AR-2087 (*Farm 2*); this is probably due to a lower Milk Yield per cow[25] in *Farm 5* because the two farms have similar mortality rates in all animal categories.

Season 2003

Figure 78: Costs of Milk Production Only (season 2003)

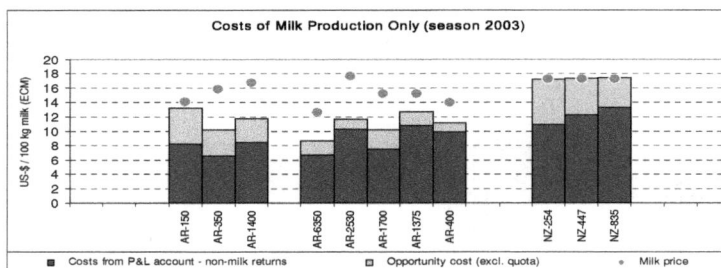

In 2003 (Figure 78), within the farms of the first group (the three typical Argentine farms), AR-350 had the lowest Costs of Milk Production Only per 100 kilograms of milk (ECM) produced. Within the second group (Farms AR-6350, AR-1700, AR-1375 and AR-400), AR-6350 (*Farm 1*) had the lowest Costs of Milk Production Only. AR-2530 or *Farm 2* (third group) had Costs of Milk Production Only that were similar to those of AR-1400.

---

[25] Milk Yield per cow and level of Non-milk Returns per kilogram of milk are related because, if two farms had similar total Non-milk returns per cow, the farm with lower Milk Yield will have higher Non-Milk Returns per kilogram of milk.

Figure 79: Non-milk Returns (season 2003)

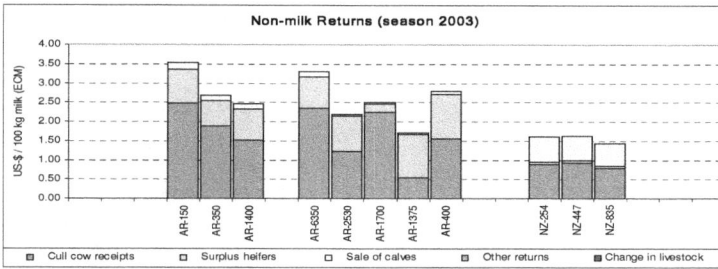

Figure 80: Animal Purchases (season 2003)

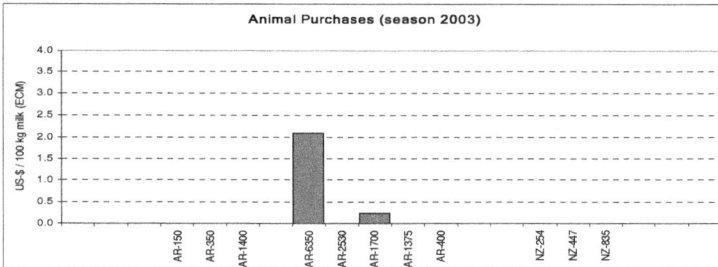

Wthin the farms of the first group (the three typical Argentine farms), AR-150 had the highest Non-milk Returns per 100 kilograms of milk (ECM) produced (see Figures 79 and 80).

Within the second group (Farms AR-6350, AR-1700, AR-1375 and AR-400), AR-6350 (*Farm 1*) again had the lowest Non-milk Returns when Animal Purchases was deducted. Therefore, AR-400 (*Farm 7*) had the highest Non-milk Returns of this group, partly due to the fact that has the lowest Milk Yield per cow of the group. AR-1375 (*Farm 5*) had lower Non-milk Returns than AR-400 and AR-1700 because it had a replacement rate of only 10%. The Replacement Rate has an impact in the analysis, because for Argentina the heifers' selling price is usually lower than the cull cows' selling price.

AR-2530 or *Farm 2* (third group) had lower Non-milk Returns than the typical farms (first group), mainly because it had lower Replacement Rates.

Adoption of New Zealand Genetics and Mortality Rates

No association was found between the number of innovations adopted by the Argentine farms and Cost of Production Only or the degree of Non-milk Returns.

However, there seems to be an association between the Mortality Rate and the adoption of New Zealand genetics (see Table 26).

**Table 26: Level of New Zealand Genetics Adopted and Mortality
and Replacement Rates (seasons 2002 and 2003)**

	AR-150	AR-350	AR-1400	*Farm 1*	*Farm 2*	*Farm 3*	*Farm 4*	*Farm 5*	*Farm 7*
Replacement Rate	26%	32%	28%	26%	21%	35%	14%	21%	20%
Mortality Rate Cows	4%	7%	6%	4%	3%	5.5%	2.5%	3%	3%
Adoption of NZ Genetics	0	0	0	0.2	1	0	1	0.67	0.3

The four farms that adopted more than 0.25 of New Zealand genetics (*Farms 2, 4, 5 and 6*) had the lowest Replacement Rates (lower than, or equal to, 21%). The same farms had the lowest Mortality Rates of cows (lower than, or equal to, 3%). It is not possible to know if the adoption of New Zealand genetics is the cause of lower Replacement and Mortality Rates, however an association between these factors was found for the Argentine farms studied in the 2002 and 2003 seasons. The following two figures illustrate this association.

**Figure 81: Level of New Zealand Genetics Adopted and Mortality
Rate (seasons 2002 and 2003)**

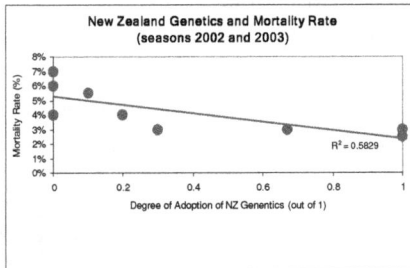

Figure 82: Level of New Zealand Genetics Adopted and
Replacement Rate (seasons 2002 and 2003)

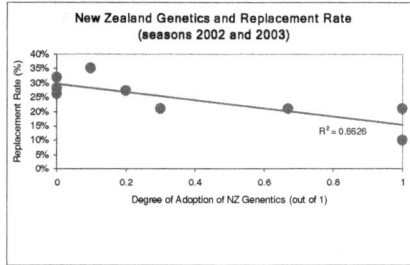

8.1.6 Costs Components (IFCN)

In the following sections, the different cost components are analysed. The components
of Total Costs defined by the IFCN are: Labour Costs, Land Costs, Capital Costs and
Costs as Means of Production (see section 4.3).

Figure 83: Costs Components as a Percentage of Total Costs
(season 2002)

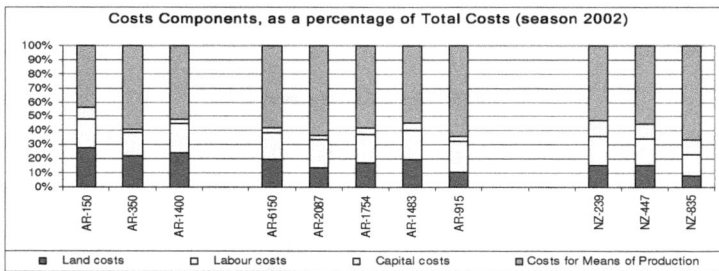

Figure 84: Cost Components as a Percentage of Total Costs
(season 2003)

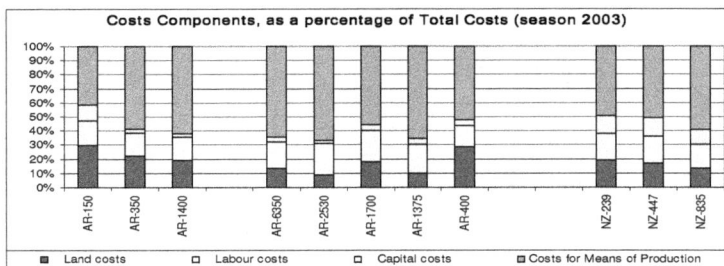

Figures 83 and 84 simply give an idea of the contribution that each of the three cost components make to the Total Costs. In general, for the New Zealand farms Capital Costs are a bigger proportion of the Total Costs.

The analysis of the Total Costs by each cost component (also called production factor) can be done independently of the analysis of other indicators and provides an idea of how each factor is utilised and how it impacts on the general performance of the business.

8.1.7 Labour

Season 2002

Figure 85: Labour Costs (season 2002)

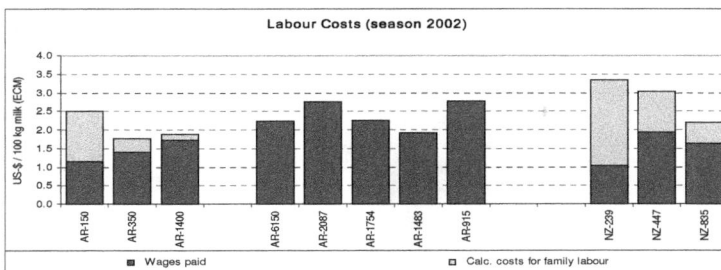

Labour Costs are the result of dividing all the labour costs (paid wages and unpaid family labour) by the kilograms of milk (ECM) produced.

In 2002 (see Figure 85), within the farms of the first group (the three typical Argentine farms), AR-350 had the lowest Labour Costs per 100 kilograms of milk (ECM) produced. Within the second group (Farms AR-6150, AR-1754 and AR-1483), AR-1483 (*Farm 4*) had the lowest Labour Costs. Within the third group AR-915 (*Farm 5*) had slightly lower Labour Costs than AR-2087 (*Farm 2*).

Figure 86: Average Wages on the Farm (season 2002)

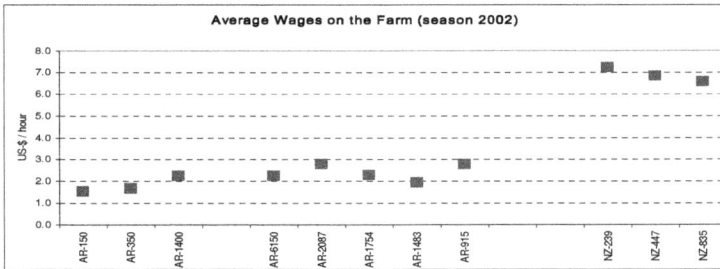

Average Wages on the Farm is the ratio between all the costs of labour and the total hours of work for the season of all the people that worked in the dairy enterprise.

Within the farms of the first group (the three typical Argentine farms), AR-1400 had the highest Average Wages on the farm. Within the second group (Farms AR-6150, AR-1754 and AR-1483), AR-1754 (*Farm 3*) had the highest Average Wages on the farm. Within the third group, AR-2087 (*Farm 2*) had slightly higher Average Wages than AR-915 (*Farm 5*).

Figure 87: Labour Productivity (season 2002)

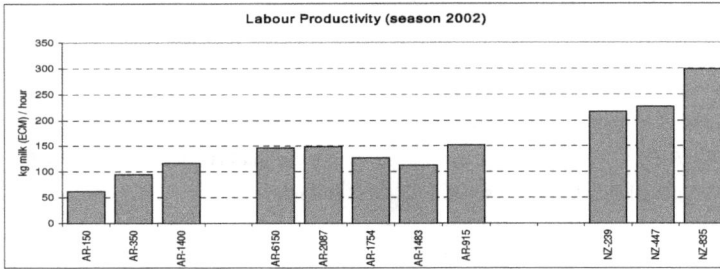

Because Labour Productivity is a physical indicator and is not affected directly by inflation, exchange rates and differences in milk prices, comparisons across farms of different groups were made. AR-915, AR-2087 and AR-6150 *(Farms 5, 2* and *1)* have the highest Labour Productivity (in that order) of all Argentine farms in season 2002. *Farm 3* (AR-1754) had slightly higher Labour Productivity than all the typical Argentine farms, and *Farm 4* (AR-1483) had slightly lower Labour Productivity than AR-1400.

Season 2003

Figure 88: Labour Costs (season 2003)

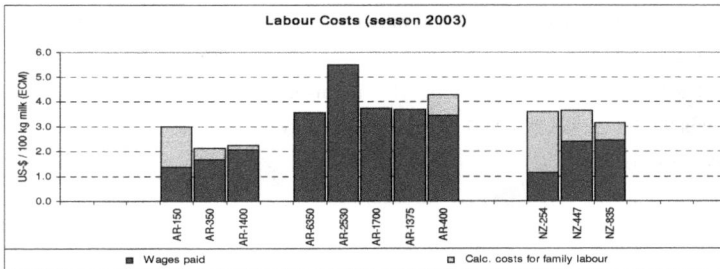

In 2003 (see Figure 88), within the farms of the first group (the three typical Argentine farms), AR-350 had the lowest Labour Costs per 100 kilograms of milk (ECM) produced. Within the second group (Farms AR-6350, AR-1700, AR-1375 and AR-400),

AR-6350 (*Farm 1*) had the lowest Labour Costs. AR-2530 or *Farm 2* (third group) had Labour Costs considerably higher than all other Argentine farms.

Figure 89: Average Wages on the Farm (season 2003)

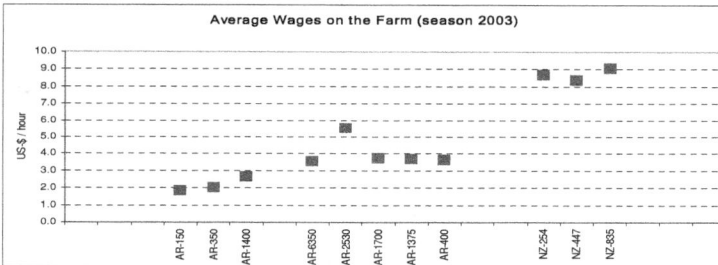

Within the farms of the first group (the three typical Argentine farms) AR-1400 had the highest Average Wages. Within the second group (Farms AR-6350, AR-1700, AR-1375 and AR-400), all of them have very similar Average Wages. AR-2530 or *Farm 2* (third group) had considerably higher Average Wages than all the other Argentine farms in this season.

Figure 90: Labour Productivity (season 2003)

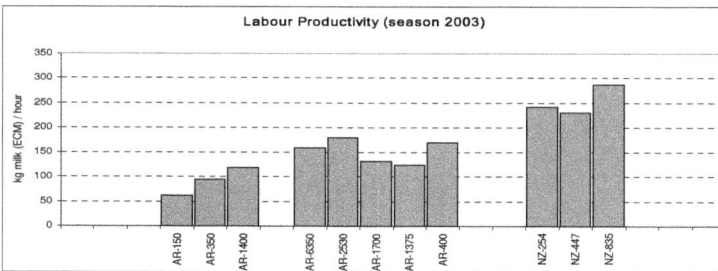

Because Labour Productivity is a physical indicator as was previously mentioned, comparisons across farms of different groups will be made. AR-2530, AR-400 and AR-6350 (*Farms 2, 7* and *1*) have the highest Labour Productivity (in that order) of all Argentine farms in season 2003. *Farms 3* and *4* (AR-1700 and AR-1375, respectively)

had slightly higher Labour Productivity than all the Argentine typical farms (*Farm 3* slightly higher than *Farm 4*).

Summary for Labour (seasons 2002 and 2003)

The wages per hour of work increased in the farms, of both countries, from one season to the other. In the typical New Zealand farms, the wages increased approximately 27%. In the typical Argentine farms, the wages increased 19%. In the Argentine case studies, the wages increased between 59% in *Farm 1*, to 106% in *Farm 4*.

Farm 2 (AR-2087/2530), paid the highest average wages per hour in both seasons, and had the second highest Labour Productivity in 2002 and the highest in 2003.

The Argentine case studies had, in general, higher Labour Productivity, higher Average Wages, and higher Labour Costs than the typical Argentine farms.

Adoption of New Zealand Innovations and Labour Productivity

The factors that affect Labour Productivity are so numerous that is very difficult to assess the impact of adopting New Zealand innovations on it. However nearly all the case studies in both seasons (with the exception of AR-1483 in 2002) had Labour Productivity higher than all the typical Argentine farms. The following figure (Figure 91) illustrates this. And interestingly, the three farms that most fully adopted the sixth innovation (*Less than 15 cows per Set of Teat-cups, etc*) had the highest Labour Productivity (average of the two seasons); these farms were *Farm 7*, *Farm 2* and *Farm 1*.

Figure 91: Number of New Zealand Innovations Adopted and Labour Productivity (seasons 2002 and 2003)

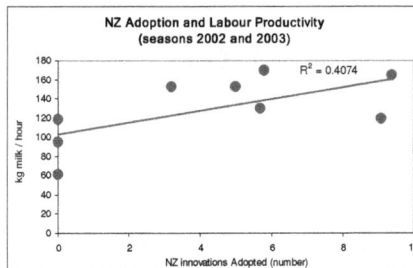

8.1.8 Land

Season 2002

Land was, on average, 77% of total assets for the typical Argentine farms in 2002. For case studies AR-6150 (*Farm 1*), AR-2087 (*Farm 2*), AR-1754 (*Farm 3*) and AR-1483 (*Farm 4*), which are the case studies that own land, land was 69%, 65%, 61% and 59% of the total assets, respectively. For the typical New Zealand farms, land was, on average, 56% of the total assets of the business.

Figure 92: Land Costs (season 2002)

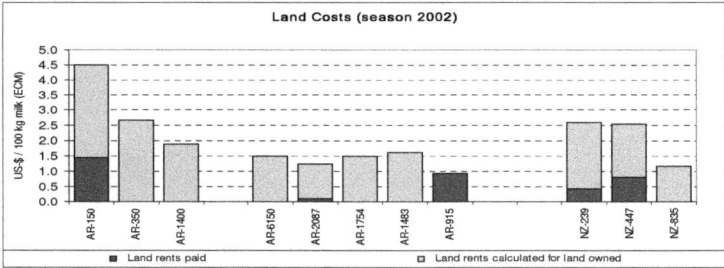

In 2002 (see Figure 92), within the farms of the first group (the three typical Argentine farms) AR-1400 had the lowest Land Costs per 100 kilograms of milk (ECM) produced. Within the second group (AR-6150, AR-1754 and AR-1483), AR-1754 (*Farm 3*) had the lowest Land Costs. Within the third group, AR-915 (*Farm 5*) had lower Land Costs than AR-2087 (*Farm 2*).

Figure 93: Market Value of Land

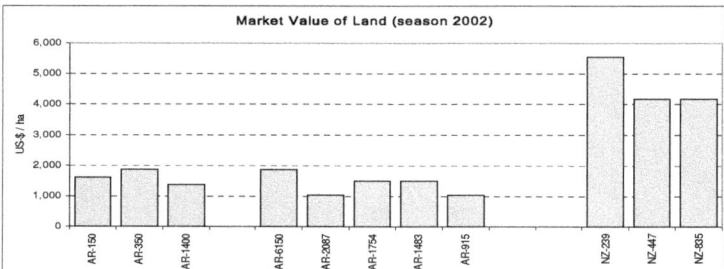

201

There are mainly four categories of Market Value of Land within the Argentine farms:

- The more valuable, and possibly the most productive land, is the land of AR-350 and *Farm 1* (AR-6150).

- This is followed, in value, by the land of AR-150.

- Then, the land of AR-1400, which is similar to *Farms 3* (AR-1754) and *Farm 4* (AR-1483).

- Lastly, the land value of *Farm 2* (AR-2087), *Farm 5* (AR-915 only in season 2002) and *Farm 7* (AR-400 only in season 2003).

Figure 94: Level of Land Rents (season 2002)

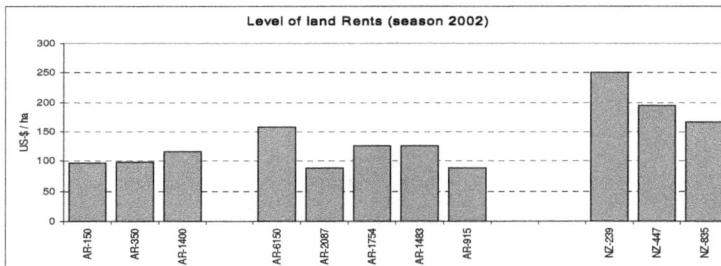

The Level of Land Rents is the value that was used as the opportunity cost for land, in the farms that did not rent land; and as a real land cost in the farms that leased land. The Level of Land Rents in 2002 was proportional to the Market Values of Land.

Figure 95: Land Productivity (season 2002)

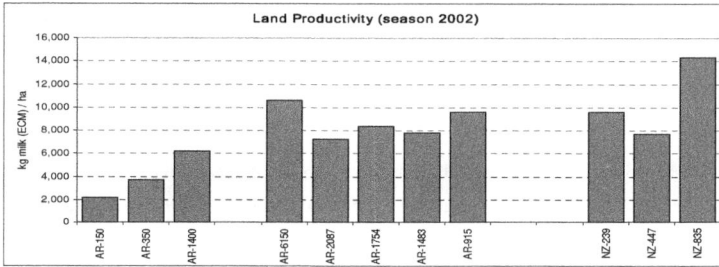

When comparing Land Productivity of different farms it is important to take into account the feeds that were brought in from other land that is not included in the analysis:

- For typical farms see section 4.4.7.

- From the case studies, *Farm 1* brought in 43% of total feeds, *Farm 2* brought in 49% of total feeds, *Farm 3* brought in 39% of total feeds, *Farm 4* brought in 55% of total feeds, *Farm 5* brought in 63% of total feeds, and *Farm 7* brought in 40% of total feeds (the proportions are approximate). The typical Argentine farms have a lower proportion of the total feeds brought in than the case studies because in the typical farms the land used to produce feeds by the cropping enterprises of the farms is included in the analysis. In the case studies, in contrast, the dairy enterprises (the *Farms*) buy the feeds from the cropping enterprises at market prices and the land used by the cropping enterprises is not included in the analysis. These differences in the methodology do not have any impact on the Land Costs; however they make it more difficult to compare Land Productivity between the Argentine farms.

On *Farm 1,* the land utilized to raise the replacement heifers is not included in the analysis because they sell the heifers calves, and then buy back the replacement heifers. This increases the Land Productivity because a category of animals that does not produce milk is not taken into account.

Due to differences in the proportion of feeds brought in, and differences in methodologies used, is not possible to accurately compare the land productivity of the farms.

Season 2003

Land was, on average, 81% of total assets for the typical Argentine farms in 2003. For case studies AR-6350 (*Farm 1*), AR-2530 (*Farm 2*), AR-1375 (*Farm 3*) and AR-1375 (*Farm 4*), which are the case studies that own land, land was 69%, 62%, 73% and 62% of the total assets, respectively. For the typical New Zealand farms, land was, on average, 56% of the total assets of the business.

Figure 96: Land Costs (season 2003)

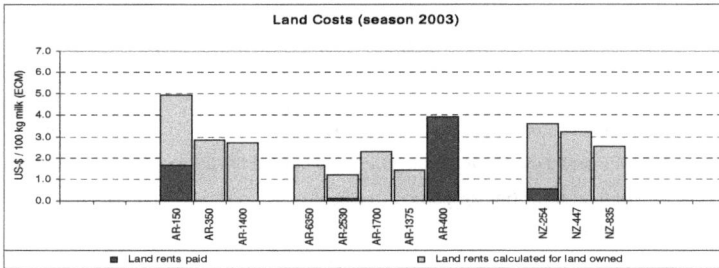

In 2003 (see Figure 96), within the farms of the first group (the three typical Argentine farms) AR-1400 had the lowest Land Costs per 100 kilograms of milk (ECM) produced. Within the second group (Farms AR-6350, AR-1700, AR-1375 and AR-400), AR-1375 (*Farm 4*) had the lowest Land Costs. AR-2530 or *Farm 2* (third group) had the lowest Land Costs of all Argentine farms.

Figure 97: Market Value of Land (season 2003)

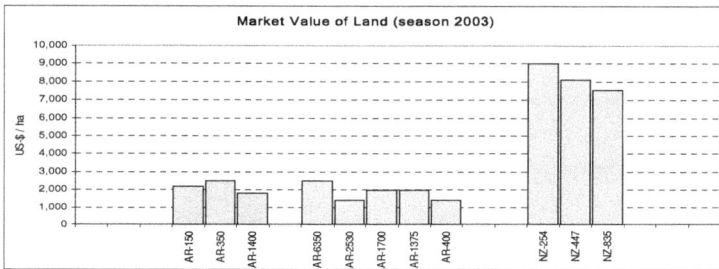

The four categories of Market Value of Land within the Argentine farms described in the 2002 season remained unchanged in 2003. However, Market Value of Land increased by 1.35 times in the Argentine farms between 2002 and 2003.

Figure 98: Level of Land Rents (season 2003)

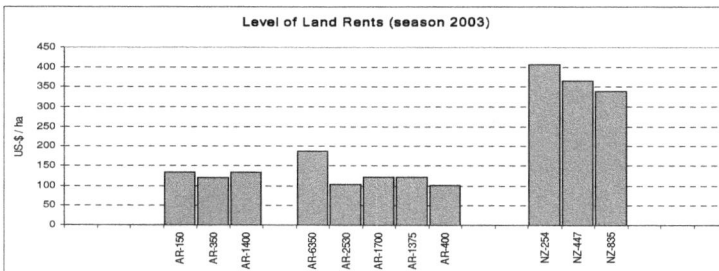

The Level of Land Rents in 2003 was proportional to the Market Values of Land, ranging from 4.8% (AR-350) to 7.7% (AR-1400, AR-6350, AR-1700, and AR-1375).

Figure 99: Land Productivity (season 2003)

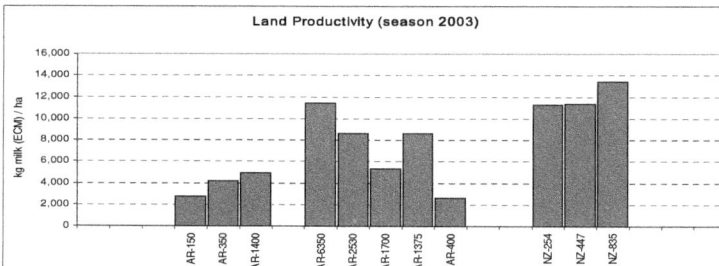

Due to differences in the proportion of feeds brought in, and differences in methodologies used, it is not possible to accurately compare the land productivity of the farms. However is important to show the Land Productivity in order to understand the Land Costs; as Land Costs (US-$ / 100 kilograms milk) are the result of Land Rents (US-$ / ha) per hectare divided by the Land Productivity (kilograms milk / ha).

Summary for Land

The case study farms had, in general (with the exception of AR-400 in 2002) lower Land Costs per kilogram of milk produced than the typical Argentine farms.

In season 2003, when comparison across groups can be made more reliably, AR-2530 (*Farm 2*) had the lowest Land Costs followed by AR-1375 (*Farm 4*) and then AR-6350 (*Farm 1*).

Adoption of New Zealand Innovations and Land Costs

The case studies, all of which have adopted the first New Zealand innovation (*Focus on Production per Hectare*), had lower Land Costs per kilogram of milk than the typical Argentine farms in both seasons (with the exception of AR-400 in 2002).

The two farms with lower Land Costs were also the two farms that adopted the highest proportion of New Zealand innovations (*Farms 2 and 4*). This does not necessary imply that the fact of adopting New Zealand innovations has an impact on Land Costs per kilogram of milk produced, however proves an association between the two factors (see Figure 101).

Figure 100: Number of New Zealand Innovations Adopted and
Land Costs (season 2003)

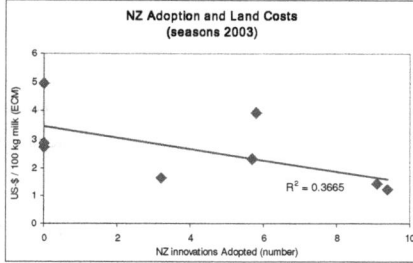

8.1.9 Capital

Figure 101: Capital Costs (land not included) (season 2002)

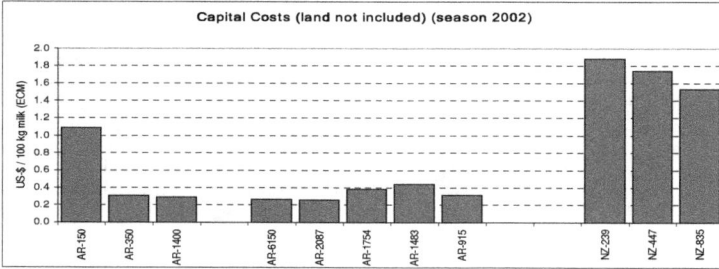

Figure 102: Capital Costs (land not included) (season 2003)

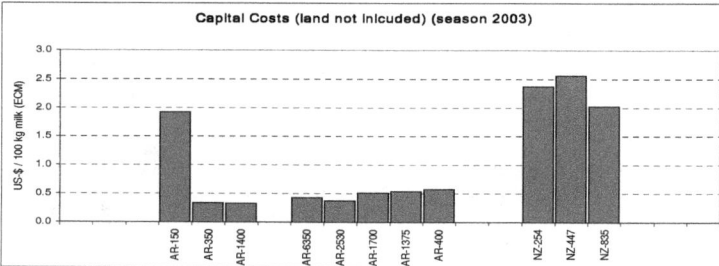

Capital Costs (on an annual basis), were calculated using a real interest rate of 6% for borrowed funds, and a real rate of 3% for owner's capital (buildings, machinery, livestock and others).

Capital Costs were similar and relatively low for most Argentine dairy farms in both seasons (see Figures 101 and 102) mainly because they did not have any long-term loans. AR-150 model has higher capital costs because it had a relatively large loan. *Farm 1* and *2* had the lowest Capital Costs in both seasons.

If the AR-150 is not considered (because it was the only Argentine farm with some long-term debt) an association was found between the adoption of New Zealand innovations and increments in capital costs.

Figure 103: Capital Input per Cow (season 2002)

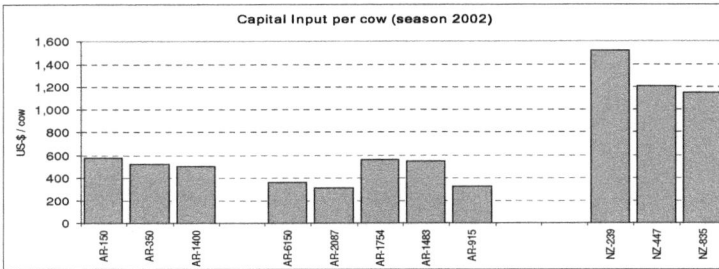

Figure 104: Capital Input per Cow (season 2003)

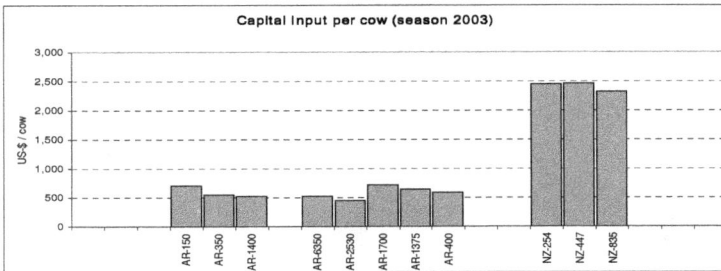

In Capital Input per Cow, only the livestock, building and machinery assets are taken into account. For New Zealand farms the co-operative shares are included in the 2003 season (in season 2002 these were added to the land). Because the value of the assets, in US-$, were not significantly affected by the inflation, comparisons across groups of farms were made for both seasons.

From the case studies, in season 2002, *Farms 1, 2* and *5* (AR-6150, AR-2087 and AR-915) were under the US-$ 400 of capital input per cow. And in season 2003, *Farm 2* (AR-2530), AR-1400, and *Farm 1* (AR-6350) had the lowest Capital Input per Cow.

Farms 3 (AR-1754/1700) and *4* (AR-1483/1375) had the highest capital input per cow of all the Argentine farms in both seasons.

Increments in adoption of New Zealand innovations were associated with higher investment in plant and machinery per cow.

Figure 105: Capital Productivity (land not included) (season 2002)

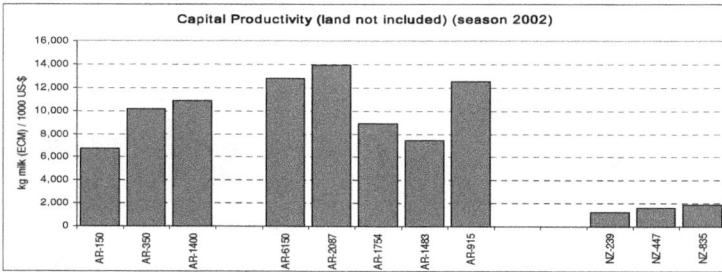

Figure 106: Capital Productivity (land not included) (season 2003)

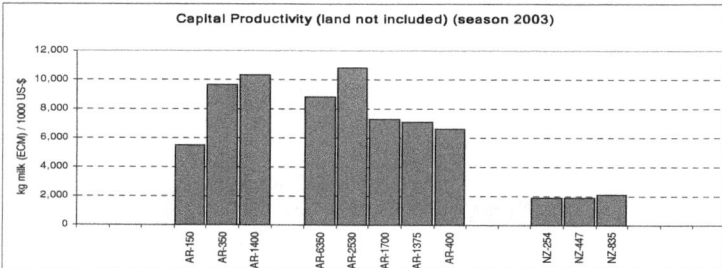

Capital Productivity is calculated by dividing the total kilograms of milk produced by the capital input (total assets minus land). In both seasons *Farm 2* had the highest Capital Productivity; followed by *Farm 1* in 2002 and by AR-1400 in 2003. The fact that for *Farm 1* the land utilised for raising heifers is not included in the analysis increases its Capital Productivity. Capital Productivity is related to Capital input per Cow and Milk Yield per Cow (Figures 104 and 105, and 108 and 109, respectively). The farms with relatively lower Capital Input per Cow and relatively high Milk Yield per Cow had higher Capital Productivity.

If the AR-150 is not considered (because it was the only Argentine farm with some long-term debt) an association was found between the adoption of New Zealand innovations and reduction in capital productivity.

8.1.10   Milk Yield

Figure 107: Milk Yield per Cow (season 2002)

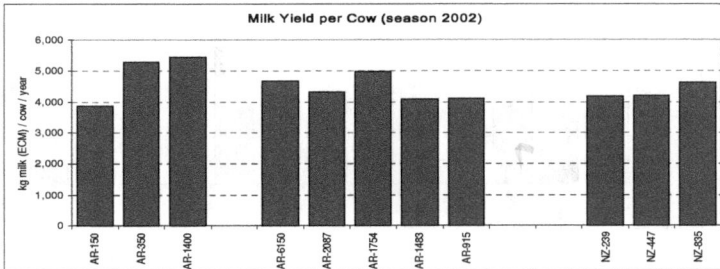

Figure 108: Milk Yield per Cow (season 2003)

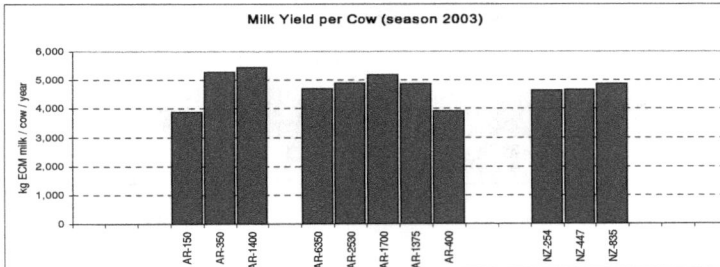

Within the six case studies, *Farm 3* (AR-1754 and AR-1700) has the highest milk yield
per cow in the two seasons analysed. Is important to mention that, on average, the cows
in *Farm 3* are the second in size (see Table 10) after the cows of *Farm 1* (AR-6150 and
AR-6350). The production per cow of *Farm 3* is higher than the three New Zealand
farms and lower than the two largest typical Argentine farms (AR-1400 and AR-350).
In general the case studies had Milk Yields per Cow that were intermediate between the
typical Argentine farms and the typical New Zealand farms.

However when the Milk Yield is calculated by kilogram of live weight of the cows (see
Table 22), *Farm 2* (AR-2087 and 2530) and *Farm 4* (AR-1483 and AR-1375) had the
highest Milk Yield in both seasons; these two farms have adopted the highest proportion
of New Zealand genetics in their herds. The following table illustrates the association
between the proportion New Zealand genetics in the farms' herds and milk yield per
kilogram of live weight.

**Table 27: Level of New Zealand Genetics Adopted and Milk Yield
per kilogram of live weight (seasons 2003)**

	AR-150	AR-350	AR-1400	*Farm 1*	*Farm 2*	*Farm 3*	*Farm 4*	*Farm 5*	*Farm 7*
Milk Yield / cow / year	3,865	5,284	5,453	4,684	4,890	5,200	4,871	4,099	3,894
Cull cow live weight	550	500	560	605	424	500	400	442	442
Milk Yield / kilograms LW	7.0	10.6	9.7	7.7	11.5	10.4	12.2	9.3	9.0
Adoption of NZ Genetics	0	0	0	0.2	1	0.1	1	0.67	0.3

8.1.11   Costs as Means of Production

Figure 109: Costs as Means of Production (season 2002)

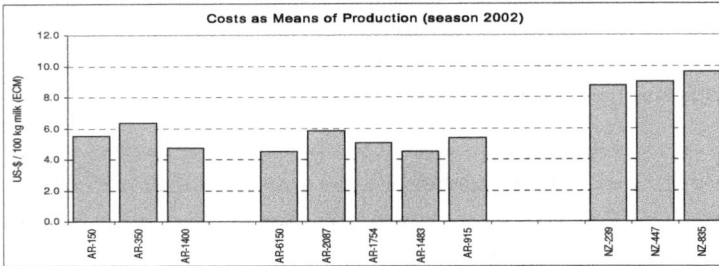

The means of production are the sum of the following expenses: animal purchases, feed expenses (purchase feed, fertiliser, seed, pesticides), machinery expenses (maintenance, depreciation, contractor), fuel, energy, lubricants and water expenses, buildings expenses (maintenance, depreciation), veterinarian and medicine expenses, insemination expenses, insurance and taxes, and other inputs.

In 2002 (see Figure 102), within the farms of the first group (the three typical Argentine farms), AR-1400 had the lowest Costs as Means of Production per 100 kilograms of milk (ECM) produced. Within the second group (Farms AR-6150, AR-1754 and AR-1483), AR-1483 (*Farm 4*) had the lowest Costs as Means of Production (slightly lower than AR-6150 or *Farm 1*). Within the third group AR-915 (*Farm 5*) had lower Costs as Means of Production than AR-2087 (*Farm 2*).

Figure 110: Costs as Means of Production (season 2003)

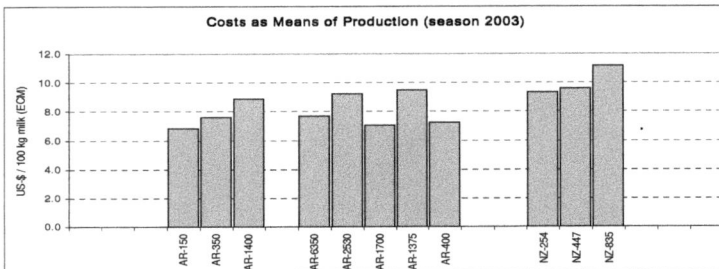

In 2003 (see Figure 110), within the farms of the first group (the three typical Argentine farms), AR-150 had the lowest Costs as Means of Production per 100 kilograms of milk (ECM) produced. Within the second group (Farms AR-6350, AR-1700, AR-1375 and AR-400), AR-1700 (*Farm 3*) had the lowest Costs as Means of Production, slightly lower than AR-400 (*Farm 7*). AR-2530 or *Farm 2* (third group) had the second highest Costs as Means of Production of all Argentine farms, after AR-1375 (*Farm 4*).

The levels of Costs as Means of Production per 100 kilograms of milk produced are difficult to analyse. In some cases (for example AR-1400 in 2002, and AR-1700 in 2003) the farms that had the lowest Costs as Means of Production were also farms that performed well financially (relatively high Entrepreneurs' Profit, Operating Profit Margin, and ROI). And in other cases (for example AR-350 in 2003, and AR-2530 also in 2003) farms with relatively high Costs as Means of Production per 100 kilograms of milk were also farms that performed well financially (relatively high Entrepreneurs' Profit, Operating Profit Margin and, ROI).

It seems that there is a slight association between the adoption of New Zealand innovations and the level of Costs as Means of Production. Becuase in 2003, the two farms with the highest Costs as Means of Production (AR-1375 and AR-2530) are also the two farms with the highest New Zealand innovations adopted; but the third farm in Costs as Means of Production was AR-1400, a typical farm (which has not adopted any of the innovations).

8.2 Chapter Summary

- The case studies with the best financial performances were *Farm 2* (AR-2087 and AR-2530), *Farm 1* (AR-6150 and AR-6350), *Farm 5* (AR-915), and *Farm 3* (AR-1754 and AR-1700). The typical Argentine farms, which performed similarly well, were AR-350 and AR-1400.

- An association between number of New Zealand innovations adopted and higher Return on Investment was found.

- No clear association was found between number of innovations adopted and the Entrepreneurs' Profit.

- Higher Operating Profit Margin seems to be associated with lower levels of adoption of New Zealand innovations.

- No clear association was found between the number of innovations adopted and the Costs of Milk Production Only.

- An association was found between the proportion of New Zealand genetics in the farms' herds and milk yield per kilogram of live weight.

- From the nine farms analysed, the four farms that more fully adopted New Zealand Genetics, had the lowest Mortality and Replacement rates in both seasons.

- The case studies had higher Labour Costs per kilogram of milk produced, higher Average Wages, and higher Labour Productivities than the typical farms in both seasons. The three farms that more fully adopted the sixth New Zealand innovation (*Less than 15 cows per Set of Teat-cups, etc*) had the highest Labour Productivity.

- The case studies, all of which have adopted the first New Zealand innovation (*Focus on Production per Hectare*), had lower Land Costs than the typical Argentine farms in both seasons. The two farms with then lowest Land Costs per kilogram of milk were also the two farms that adopted the highest proportion of New Zealand innovations.

- If the AR-150 is not considered (because it was the only Argentine farm with some long-term debt) an association was found between the adoption of New Zealand innovations and increments in capital costs, and also with reduction in capital productivity. Additionally, increments in adoption of New Zealand innovations were associated with higher investment in plant and machinery per cow.

- A slight association was found between the adoption of New Zealand innovations and higher Costs as Means of Production.

9   DISCUSSION

9.1   Competitiveness of Argentine Dairy Farms

In this section the competitiveness[26] of Argentine dairy farms is addressed by discussing the results from the comparison of typical and real Argentine farms with New Zealand typical farms.

New Zealand dairy farms were considered traditionally the most competitive in milk production costs. However milk production costs in 2002 and 2003 on the Argentine dairy farms (typical and case study farms) were lower than on the New Zealand typical farms. This was mostly explained by two factors: a) Argentine farms had lower labour costs; and b) Argentine farms had lower capital costs per kilogram of milk produced than New Zealand farms.

In 2002, and possibly due to the devaluation of the Argentine currency, the typical Argentine farms became the most competitive in milk production costs (in US-$) of all the typical farms in the main milk production countries in the world (IFCN, 2003).In 2003 the same happened (IFCN, 2004). This implies that in Argentina, during 2002 and 2003, farmers produced milk at a lower milk price than in any other country in the world. The exchange rate of the Argentine currency averaged AR-$ 3 for each US-$ since March 2003 (calculations based on *FXHistory: Historical currency exchange rates*, 2005) and it is expected to continue at similar levels in 2005 (BCRA, 2005). Therefore, Argentina will continue to produce milk at competitive costs during 2004 and will probably continue to do so in 2005. However, the 8% inflation during 2004 (calculations based on the Wholesale Price Index INDEC, 2005) and the 5% to 8% inflation expected for 2005, could erode this advantage.

This competitiveness, together with raise in international dairy products prices, increased the export possibilities of Argentine dairy companies; during 2004 they exported more than 20% of the total milk production, which was the second highest level for the last 15 years (La Opinión, 2004).

---

[26] *Competitiveness* is defined "as the ability to profitably create and deliver value at prices equal to or lower than those offered by other sellers in a specific market" (Harrison & Kennedy, 1997, p.16).

However, the fact that the Argentine dairy farms (typical and case studies) were able to produce milk at lower costs than the typical New Zealand farms in 2002 and 2003, does not necessarily mean that they had higher profitability than the typical New Zealand farms. On the contrary, the New Zealand typical farms had higher Entrepreneur's Profit per kilogram of milk produced than the typical Argentine farm at a similar management level and relatively similar milk prices (for example comparing AR-150 in 2002 with NZ-239 in 2003 or AR-150 in 2003 with NZ-254 in 2002, see Figures 10 and 11); and the Return on Investment (ROI) of these Argentine and New Zealand farms (see Figures 14 and 15).

The similar ROI of two comparable farms in different countries could probably be explained by the fact that market values of assets tend to stabilize at levels at which returns are acceptable. In the case of the typical dairy farms in New Zealand and Argentina, in which land is the main asset, this means that the market value of land will tend to stabilize at values at which the returns of dairy farming are acceptable.

The advantage in labour costs of the typical Argentine farms was mainly due to much lower average wages per hour of work than in the typical New Zealand farms (average wages were between 3 and 4 times higher in New Zealand farms). At the same time it was found that the New Zealand typical dairy farms had higher investment in plant and machinery per cow (investment in plant and machinery between 2 and 3 times higher in New Zealand farms), and much higher labour productivity than the typical Argentine farms (labour productivity approximately 3 times higher in New Zealand farms). Therefore, the relationship between investment in plant and machinery and the cost per hour of labour is probably the underlying force that has induced the New Zealand typical farms have relatively high labour productivities, and made Argentine farms not so focused on it.

The advantage in capital costs of the typical Argentine farms was mainly explained by the lower level of debt utilized by them, and to a lesser degree by the lower investment in plant and machinery per cow than the New Zealand typical farms. Argentine farms, because they do not usually utilize long-term loans, have lower annual interest paid on liabilities than New Zealand farms. Argentine farmers do not usually utilize any kind of long-term financing because access to credit is usually limited and because interest rates are relatively high. In marked contrast, in New Zealand farms, long-term loans are utilized frequently to finance the farm business. Relatively easy access to credit at affordable rates is an advantage that New Zealand farmers have and that Argentine farmers do not have. Concluding, the lower capital costs of the typical Argentine dairy farms in 2002 and 2003 helped them to produce milk at the lowest cost. However Argentine dairy farmers have an important disadvantage compared to New Zealand dairy farmers; they are not able to utilize higher levels of debt.

The Argentine typical farms had lower labour and capital costs, however the New Zealand typical farms had lower land costs than the Argentine typical farms. The latter is true despite the fact that land is more expensive in New Zealand than in Argentina, because the New Zealand typical farms achieved much higher land productivity.

The following section discusses how the adoption of New Zealand innovations can assist Argentine dairy farmers to increase their competitiveness, especially in respect to labour productivity and land costs.

9.2   Benefits of Adopting New Zealand Principles

In this section the research question is directly addressed: *Can Argentine dairy farmers benefit by adopting New Zealand dairy farm ideas, practices and technologies?*

The present study has shown that the answer to this question is: Yes, Argentine dairy farmers can benefit from adopting New Zealand innovations in many ways. These are discussed briefly below in three sections: The first section summarizes the main consequences, advantages, disadvantages and constraints of adopting New Zealand innovations proposed by the case study farms and the dairy sector experts interviewed. The second section addresses the association between level of adoption and increments in ROI; and the third section discusses the association between labour and New Zealand innovations.

9.2.1 Main Consequences, Advantages, Disadvantages and Constraints of Adopting NZ Innovations

1) Focus on Production per Hectare

All of the seven case studies adopted this New Zealand innovation.

The main reason explained by the Argentine case study farmers for focusing on production per hectare is because land is their most limiting production factor. As one of the farmers mentioned that capital and skilled labour are difficult to obtain, however they are still easier to obtain than land. Therefore maximizing production per hectare, Argentine farmers are trying to better utilise their most limiting production factor, land.

The two case studies that leased land also believed in the importance of maximizing production per hectare, because rent was their main fixed cost and they have to dilute it by producing the greatest volume of milk possible.

2) Give Marked Importance to Pasture Production

All except one of the case studies stated that pasture was their cheapest feed; therefore it was important for them to maximize its production and utilization.

Level of fertilizer application was an indicator of the focus of the case study farms on maximizing pasture production. Two of the farmers mentioned that Argentine experts recommend increasing phosphate levels of soils up to 25 ppm for lucerne and up to 15 ppm for fescue. Four of the seven case studies were trying to reach the recommended phosphate levels on their soils. The two case studies that leased the land were fertilizing very little because they did not want to invest in land which they did not own, as they were not sure whether they would renew the leasing contract with the landowner.

The adoption of this principle was related to the relative importance of pasture in the total cows' annual requirements. For example, one of the farmers mentioned that grazed grass was only 28% of the total feeding requirements of his farm and consequently, he was not focused on pasture production whereas in another of the farms, for which grass was the main feed, the farmer was focused on pasture production because grass was the main driver of increments on production per hectare.

During the interviews one consultant and industry expert, mentioned that nowadays most consultants and top farmers know that certain pastures need high phosphate levels in soils and have high responses to nitrogen fertilizer. In addition, other consultants and farmers noted the beneficial effects of higher quantities of phosphate and nitrogen fertilizers on Argentine dairy farms.

3) Quantitative Pasture Monitoring

Three out of seven of the case study farms had fully adopted this New Zealand innovation.

Nowadays in Argentina, many farmers and consultants are interested in pasture monitoring. Further, some consultants are offering to implement pasture monitoring recording programmes on dairy farms (researcher's experience in Argentine dairy industry).

Three of the case studies had a formal pasture-monitoring programme in place. It was mentioned that in order to manage pasture efficiently, constant monitoring is essential. In one of the farms, the farmer monitors the pasture cover and growth frequently, especially in key moments of the year, in order to regulate the quantity of concentrates that is being fed.

Another farmer mentioned that he finds that the adoption of this innovation has several advantages: it helps them to better manage the cows' supplementation, it provides useful information about the stock of grass available to be eaten in order to notice in advance feed deficits (they calculate a ratio of the kilograms of dry matter available per milking cow); and in some circumstances when it is necessary, it provides them with useful data to redistribute hectares across the dairy farms of the company.

In two of the case study farms, the person that was doing the monitoring was not the dairy farmer, consequently the person that was monitoring, was not the person who was deciding which paddock would be grazed. In typical New Zealand farms the same person usually does both things. This difference could be quite significant, because information and time are lost in the process of reporting the results of pasture monitoring to the decision-maker.

In contrast, one of the farmers mentioned that he was not doing quantitative pasture monitoring because the importance of pasture for his production system was not so high as to justify the expenditure of time on pasture monitoring. Therefore, the adoption of quantitative pasture monitoring is likely to be associated with the proportion of grazed grass in the total annual requirements of the cows.

Two other farmers considered that they could manage the grazing of the cows without the need of formal monitoring. These two farms were the smallest of all the case studies and both farmers had experience in dairy farming. Therefore, experienced farmers managing relatively small farms found the adoption of quantitative pasture monitoring not very necessary nor advantageous.

4) Utilization of Formal Pasture Budgets

This innovation is closely related to the previous one, the same three case study farms that had fully adopted *Quantitative Pasture Monitoring,* had adopted also this New Zealand innovation.

Pasture budgets are useful for detecting feed deficits and surplus, they are utilized to decide when to buy feedstuffs, when to fertilise, or when to harvest grass surpluses. Additionally, one of the case study farmers mentioned that he finds pasture budgets useful to detect differences in pasture production or utilization between the different dairy farms of the company. Another farmer mentioned that he usually uses pasture budgets to define stocking rates in the different farms that he manages.

It is the opinion of the researcher that, without utilizing pasture budgets, it is difficult to plan and to manage feeding in a dairy farm. Pasture budgets help farmers to better utilize their grass, which generally represents 50% of the annual feed requirements of the cows in Argentine dairy farms.

5) Skilled and Motivated People Working on Farms

Four out of seven of the case study farms had fully adopted this New Zealand innovation.

The main advantage mentioned of this innovation was the fact that educated, skilful, and motivated people are able to manage the owner's farm (when the owner himself is not operating it) without the need of constant supervision. Additionally more motivated and educated people are usually more proactive in constantly searching for opportunities and changes that can increase efficiency and profitability. Furthermore, as most dairy farm owners in Argentina are relatively high educated (finished secondary school and very often with a tertiary or university education), having educated people operating the dairy farms facilitates communication between them and the people who operate their dairy farms.

It was found in the present study that the main reason why more motivated and educated people do not operate Argentine dairy farms is because working and job conditions in Argentine farms are not good enough to attract them.

Some of the case study farmers mentioned that they could see some advantages in having more skilled and motivated people working in their dairy farms. However, they mentioned that they were not sure whether the benefits were higher than the costs. They believe that the transition, from traditional Argentine dairy farming operating structure to something similar to the New Zealand "sharemilking" structure, would be very costly. A big investment would have to be made in buildings and machinery in order to create an environment that would attract and retain educated and trained people.

Two couples, of what could be some of the first "Argentine sharemilkers", were also interviewed. They mentioned some reasons that constrain the employment of more educated people on dairy farms. Firstly, they mentioned that most Argentine landowners are not willing to invest in buildings and machinery in order to improve the working conditions on their dairy farms. Secondly, they mentioned that Argentine landowners are probably not interested in changing an operating structure that is currently working fine for them. Thirdly, they mentioned that Argentine landowners think that if they employ educated people on their farms they would lose part of the control over their dairy farms.

In contrast, dairy farm owners mentioned that is difficult to find young, skilled and motivated people to work on dairy farms because young people do not want to work in the countryside and because dairy farming is seen as a low-status activity. They also mentioned that very often young people who were raised on dairy farms and who had then achieved a tertiary or university education, are not willing to return to dairy farming. A disadvantage that farm owners found in highly educated young people was that they are not accustomed to the working conditions of a dairy farm, even when there has been an investment in buildings and machinery. Consequently they usually work in dairy farming for a limited period of time to save money and then they look for a less demanding job.

Both of the "Argentine sharemilkers" couple also mentioned that they agree with the perception of some of the case study farmers, that to employ young university graduates, who do not have experience in dairy farming, is very costly. However, all four of them mentioned that, if in the future they had the opportunity to be the owners of a dairy farm they would try to have skilful and motivated people managing their dairy farms in the appropriate working conditions, because they believe that the extra benefits would be bigger than the extra costs.

One important disadvantage that the "sharemilkers" found in Argentina in contrast to New Zealand was the limited access to credit. This restricts their capacity to buy cows and machinery in order to firstly rent a farm and then evolve towards farm ownership.

6) Less than 15 cows per Set of Teat-cups, and Other Innovations that Impact on Labour Productivity

This was the New Zealand innovation less adopted. The farm that adopted it the most had 20 cows per set of teat-cups.

The number of cows milked per unit of teat-cups was found to be relatively constant with each country, but markedly different between countries. Typical New Zealand farms milked 15 cows per set of teat-cups, and Argentine farms (the three typical and the seven case studies) milked around 25 cows per set of teat-cups. This means that in Argentine typical farmers would have 40 units of teat-cups to milk 1000 cows, in New Zealand in contrast, typical farmers would have 67 units to milk the same number of cows. This ratio (cows per set of teat-cups) is crucial for the organization of labour and has a direct impact on labour productivity (calculated as milk produced per hour of work). An important part of the difference in labour productivity between New Zealand and Argentine typical farms (labour productivity on New Zealand farms was nearly 3 higher than on Argentine farms) is related to the fact that Argentine dairy farms milk 1.7 times more cows per set of teat-cups than New Zealand farms. This means that milking must take much longer on Argentine farms probably by about 1.7 times, because of the smaller number of teat-cups. Other causes for the difference in labour productivity are probably the utilization of other practical technologies, motivation of the staff, infrastructure of the country, and skilfulness of the staff.

In the present study, an association was found between labour productivity and number of New Zealand innovations adopted (see Figure 91). Further, the three farms that most fully adopted the sixth innovation (less than 15 cows per set of teat-cups, etc) had the highest labour productivity (average of the two seasons).

The main advantage found was that, at constant average wages, increments in labour productivity reduces labour costs per kilograms of milk produced, and can therefore reduce milk production costs.

One of the farmers stated that the advantage that he found in increasing labour productivity was that he was able to pay better wages per person without increasing total labour costs. By offering better wages, he was also able to have access to more educated people.

The number of cows milked per set of teat-cups is related to the duration of each milking, which is related to the time that people and cows spend in the milking shed at each milking. In New Zealand is traditional to try not to milk for more than four hours per day; probably because a person becomes tired after more than two continuous hours of an activity that is relatively demanding both physically and mentally. Furthermore, the cows need to be grazing on the paddocks for as long as possible in order to eat all the pasture they require. Another problem of milking too many cows per set of teat-cups is that farmers do not have enough time to do other activities on the farm, and also that because cows spend too much time in the collection yard, they leave their urine and faeces there instead of on the paddocks.

7) Seasonal Calving, One or Two Calving Periods per Year

Five out of seven of the case study farms had fully adopted this New Zealand innovation.

The main advantages of seasonal calving mentioned by the case studies are the following:

223

- Events occur in sequence, and consequently only one important task must be done at the same time (calving, rearing of calves, mating and drying off). This enables the farmer to adopt a more orderly and planned approach to mating. For example, to plan mating in advance, to do some special training or to learn new advances in mating, then execute the plan, and finally measure results.

- Most of the cows have common requirements at a given time, which increases the possibility of feeding them better in those periods when they have higher requirements (peak of lactation and mating).

- Seasonal calving makes it possible to produce a higher proportion of milk during the periods of the year when the milk price is higher.

- In hot areas of the country, seasonal calving makes it possible to avoid the calving of cows during summer, which can reduce the mortality rate of cows and calves during calving and heat stress during lactation.

- Seasonal calving facilitates the cleaning, repair and maintenance of the dairy farm plant, machinery, buildings and general structure.

- Seasonal calving facilitates the organization of holidays for the staff and limits the stress to only specific times in the year. It is important to mention that three out of the four Argentine "sharemilkers" interviewed ranked seasonal calving as one of the most important New Zealand innovations. Therefore, seasonal calving could be related to the possibility of attracting more educated people.

- Seasonal calving usually results in a decrease in the age of heifers at first calving, because the fact that all the heifers are similar in age makes it possible to rear them in a more simple and orderly fashion.

- As one of the farms stated: *'everything is more simple if it is seasonal'*.

The main disadvantages of seasonal calving mentioned by the case studies are the following:

- The need for labour is seasonal, and this could complicate the management of human resources on some occasions.

224

- Often larger milking sheds are needed in order to cope with the milking of a larger number of cows, and with higher milk yield during peak lactation. In contrast, in all year around calving systems the number of cows and their milk yields are relatively constant during the whole year. There is also the need for more space to rear calves, however that space would only used for part of the year.

- One of the farmers stated that although the highest milk prices are usually paid during winter, in some seasons that is not the case and consequently he finds it risky to have a big proportion of milk produced in a short period of time. One expert added that in places with relatively unstable weather conditions it could also be risky to have all the cows calving at the same time. These two risks could be partially solved by having two calving seasons.

- The fact that Argentine Holstein cows calve on average every 13 months (and not 12 months) could be a constraint to the adoption of seasonal calving, especially if farmers do not want to adopt New Zealand genetics. However there are some Argentine farms that adopted seasonal calving systems without the adoption of New Zealand genetics: these farms have two calving seasons per year, and very often each cow calves every 18 months.

Reasons for having **two calving seasons** in the same dairy farm or in two different dairy farms of the same company:

- Probably the most important cause is financial management, especially cash flows, many expenses are all year round and farmers do not want to be forced to use short-term loans. This is especially relevant for farms that lease land because, added to all the other expenses, they have to pay the rent.

- Another important reason could be utilization of the empty cows. Every farm has a percentage of empty cows that are sometimes useful cows that, for some reason, did not conceived. In Argentina, prices for cull cows and heifers are usually not very strong and farmers try to keep good cows and heifers if they can, rather than selling them.

- In Argentina, during the last months of summer, the milk price traditionally increases. Therefore a second calving season in late spring or early summer can be beneficial in order to capture those months of relatively high milk prices.

- The same company could have two dairy farms in different regions. One of the regions could have weather conditions that make it more suitable for dairy farms with spring calving, and the other could be more suitable for autumn calving.

Main constraints for seasonal calving in Argentina:

- The main limiting factor for the adoption of seasonal calving in Argentina proposed by the case farmers is the tacit opposition of the dairy companies. There are some dairy farmers who produce milk seasonally and they have not had any problem with the dairy industries so far. However, dairy companies are unpredictable and therefore the consequences of adopting seasonal systems are unknown.

- Another constraint of seasonal calving is the financial management (especially cash flows), which is very difficult when most of the milk is produced in a short period of time.

- The adoption of two calving seasons could solve both of these constraints. But, at the same time, some of the advantages of seasonal calving would be reduced.

8) New Zealand Genetics

Type and breed of cow is an ongoing debate among dairy farmers in Argentina. Many farmers believe what one of the experts stated: that there is some evidence that supports the use of big (and usually North American genetics cows) in Argentine farms, but other evidence supports the opposite.

Some **advantages** of New Zealand genetics mentioned by the case studies are the following:

- New Zealand genetics cows are selected in pastoral systems and are more prepared to walk long distances than Argentine Holstein cows.

- New Zealand genetics cows are able to calve every 12 months and therefore enable farmers to have seasonal systems. One of the Argentine farmers stated that it is not possible to have seasonal systems with North American cows.

- Another farmer mentioned that he likes New Zealand genetics cows because of *'their size, the strength of their legs and their udders'*.

- Other case study farmer mentioned that New Zealand genetics have been proved to be able to be capable of being milked, especially the New Zealand Jersey.

- One of the farmers found that the crossing of New Zealand Jersey with North American Holstein reduced the mortality at calving and calving difficulties. This also had a general positive impact on the reproductive performance of the herd. Additionally, crossbreds (Argentine Holstein and New Zealand Jersey) stayed longer in the herd because they had fewer fertility problems, fewer abortions, lameness, and mastitis. Furthermore, crossbreds showed better temperament during the milking, and they were lighter than Argentine Holstein cows and consequently caused less damage to pastures during wet weather.

- The dairy company to which one of the farms sold its milk, preferred a higher proportion of milksolids in milk because it costs less per kilogram of milksolids to transport and because it was easier to process. Therefore this company would prefer milk from crossbred cows.

Some **disadvantages** of New Zealand genetics mentioned by the case studies are the following:

- In one of the farms New Zealand Holstein Friesian genetics were used on Argentine Holstein cows for a number of years. But then they decided not to use them anymore because the herd had udder and calving problems. In addition, they did not like the appearance of the New Zealand Holstein Friesian. Interestingly, one of the Argentine dairy production experts interviewed mentioned the same two reasons as disadvantages of the New Zealand Holstein Friesian cows.

- A disadvantage of New Zealand genetics is that bull calves, especially Jerseys, are very difficult to sell in the Argentine market; and there is no market for cows and heifers of New Zealand genetics. However, it is important to take into

227

account that the selling of calves comprised less than 1% of the total returns in the typical Argentine dairy farms, and that the selling of heifers was less than 5% of the total returns. Additionally, many surplus heifers and cows are sold in Argentina as beef to slaughterhouses and not for their genetics.

- Another disadvantage mentioned was the fact that, in general, milk form New Zealand cows has a higher proportion of milk-fat than the Argentine Holstein cows, and in Argentina some dairy companies pay mainly for milk-protein. Additionally, milk-fat is energetically more expensive to produce than milk-protein. It was also mentioned that in the future this trend towards preferring milk-protein than milk-fat would continue. This is an interesting point and further research should be done in order to measure the impact of the relation fat/ protein in the milk payout. Within the New Zealand genetics, Holstein Friesian cows have a fat/ protein ratio more similar to American Holstein than the New Zealand Jersey. In New Zealand breeding schemes, the breeding companies are already putting a much higher economic value on milk protein because the New Zealand dairy company-payment scheme includes a higher payment for protein, a trend that will continue (Colin Holmes, personal communication).

- Another disadvantage of New Zealand genetics in Argentina is that New Zealand semen is more expensive and more difficult to buy because it has to be imported, and because New Zealand genetics companies do not offer many services in Argentina.

Some additional points:

- It was mentioned that North American genetics (which have a big influence in Argentine Holstein cows) are selected from a bigger population than in New Zealand and have been selected for a larger number of years. The researcher does not know if there are studies that support this statement.

- However, the present study found an association between the adoption of New Zealand innovations and lower mortality and replacement rates. Additionally, the present study found an association between the adoption of New Zealand genetics and higher milk yield per kilogram of live weight. These associations suggest that New Zealand genetics do offer some real advantages for Argentine dairy herds.

In conclusion another New Zealand farming principle should be mentioned: *the cow should fit the system.* Firstly farmers have to decide how they want to produce milk, and then find a cow suitable for their system. It seems that Argentine farmers that want to have seasonal systems would need some degree of New Zealand genetics. Additionally present research results suggest that New Zealand genetics cows are better suited to typical Argentine dairy systems, in which grazed pasture covers approximately 50% of the total annual requirements of dairy animals. However, further research should be done in order to assess which is the optimum type of cow for typical Argentine dairy systems.

9) Rearing of Calves in Groups

Six out of seven of the case study farms had adopted this New Zealand innovation.

Some consensus was found, during the interviews, around the fact that rearing calves in groups with calf-teats may not be the best system for everyone. There seem to be no doubt that this kind of calf-rearing system saves time and can have excellent results. However, if the person in charge of the rearing does not have the capacity to attend to each calf individually, despite the fact that they are in groups, the mortality of calves can be relatively high.

It also seems that the transition from the traditional calf-rearing system to raising calves in groups can be difficult; for example in one of the farms, the first time this group calf-rearing system was tried, very high mortality ocurred. However they are convinced of the benefits of this system and they are considering trying it again.

Main advantages mentioned by the case studies were:

- Less time required in order to rear the same number of calves.

- Enables the person in charge to devote more time to observe the calves, to see if they are sick and to touch their navels to assess if they are infected. In the individual-rearing system, the person is so busy warming the milk, filling the buckets and moving the stakes, that he or she does not have time to observe the calves.

- It allows the calves to socialize earlier and to get accustomed to belonging to a group.

- Calves can move freely and they can go to shade if it is too hot, or shelter from the wind when it is too cold.

- It seems that this system can be more enjoyable for the person in charge and is also more "animal friendly".

10) Style of Milking Shed and Milking System

All the case studies adopted this New Zealand innovation. Many of them are changing from more traditional Argentine milking sheds and systems to the typical New Zealand herringbones. The main cause for this widespread adoption is that these systems allow more cows to be milked per set of teat-cups and more teat-cups to be handled per person than in the traditional Argentine systems. Both of these combine to enable the staff to milk more cows per person and per hour in the New Zealand herringbones. Another advantage is that they require less capital input per set of teat-cups than in the other systems. At the same time, the New Zealand style milking sheds and milking systems, if correctly built, can be very comfortable to work in and easy to clean.

9.2.2 Return on Investment, Land, and Adoption of NZ Innovations

The adoption of the New Zealand innovations was found to be associated with increments in Return on Investment (ROI) and decreases in land costs per kilogram of milk. Additionally case study farmers mentioned that the adoption of the New Zealand innovations could increase land productivity (milk produced per unit of capital invested in land).

The New Zealand innovations related to pasture production and pasture utilization (increments of phosphate levels in soils, utilization of quantitative pasture monitoring, and utilization of pasture budgets) were considered to be most strongly associated with higher land productivity by the case study farmers.

Grazed pasture was one of the main sources of feed on the Argentine dairy farms. Pasture covered between 30% and 68% of the total cows' annual requirements on the case study farms, and between 52% and 61% of the total cows' annual requirements on the typical Argentine farms. In typical New Zealand farms, grazed pasture covered between 70% and 82% of the total annual requirements of cows. The adoption of New Zealand innovations were considered by the case studies to be related to increments in pasture production and utilization per hectare, and consequently in overall land productivity overall.

Higher land productivity was probably the main reason why the adopters of New Zealand innovations had lower land costs (capital invested in land per kilogram of milk produced).

Additionally land was the main investment for the typical and real Argentine dairy farms; this partly explains why improvements in land productivity were associated to increments in ROI.

ROI is possibly the ultimate indicator of financial performance. ROI is calculated as the operating profit (called Economic Farm Surplus in New Zealand) of a business as a percentage of the total investment in the business. Dairy farms with higher ROI provide to their owners a higher profit per dollar invested in the business. ROI is also useful to compare the returns from the investment in a dairy farm with the potential returns from other possible businesses.

9.2.3 Labour and Adoption of NZ Innovations

The adoption of New Zealand innovations was found to be also associated with increments in labour productivity. The main New Zealand principle associated to labour productivity was having less than 15 cows milked per set of teat-cups (or more than 67 teat-cups for every 1000 cows to be milked). Other innovations related to improvements in labour productivity were seasonal calving, rearing of calves in groups, and New Zealand style milking sheds and systems. Despite their higher labour productivities, farms that adopted New Zealand innovations had higher labour costs per kilogram of milk produced because they also paid higher wages, and the difference in wages was larger than the difference in labour productivity.

It is important to mention that none of the case studies fully adopted the New Zealand innovation in relation to the number of cows per set of teat-cups. It is possible that some of the case studies that had adopted many of the New Zealand innovations, including more skilful and motivated people, would have had higher labour productivities if the number of cows per set of teat-cups had been decreased. This increase in labour productivity would have been associated with lower labour costs. However, some capital invested have been required in order to increase the number of sets of teat-cups, therefore capital costs would have increased.

Further research should be done in order to study the trade off between level of investment in plant and machinery and the amount of labour needed for Argentine dairy farms. There is probably a level at which typical Argentine farms would decrease their labour costs at a rate which is higher than the rate of increment in capital costs. It is the opinion of the researcher that the costs of adding sets of teat-cups (up to certain level) to Argentine dairy milking sheds, could be repaid by lower labour costs.

9.2.4 Capital (land not included) and Adoption of NZ Innovations

An association was found between the adoption of New Zealand innovations and increments in capital costs (land costs not included), and also with a reduction in capital productivity. Additionally, increments in adoption of New Zealand innovations were associated with higher investment in plant and machinery per cow. Further research should be done in order find the explanations of these associations.

9.2.5 Means of Production and Adoption of NZ Innovations

A slight association was found between the adoption of New Zealand innovations and higher "costs as means of production"[27]. The two farms that adopted the higher level of New Zealand innovations had the highest costs of as means of production. Additionally, for these two farms costs as means of production represented the largest proportion of total costs. On the other hand these two farms had the lowest land costs of all the farms, in absolute terms and also as proportion of total costs. Further research should be done in order to better understand this association.

9.3 Non-diffusion of New Zealand innovations in Argentina

Many Argentine farmers that have been adopting some of the New Zealand principles and practices on their farms, and many New Zealand consultants and researchers that have travelled to Argentina to communicate the New Zealand dairy farming principles, share the following question that motivated the present study: *Why is it that New Zealand practices and principles, despite the fact that they appear to be so beneficial to Argentine farms, have not been widely adopted in Argentina?*

The Theory of Diffusion (reviewed in Chapter 3) stated that the four main factors that influence the spread of innovations are: a) the perceived attributes of innovations, b) the communication channels, c) the characteristics of the social system, d) and the extent of change agents' promotion efforts. From these factors "the perceived attributes of innovations" is the most important in influencing diffusion. The attributes are: relative advantage, compatibility, complexity, trialability and observability (Rogers, 2003). From these attributes, "perceived relative advantage", also called "perceived potential benefits", was suggested by the literature to be the most important in influencing adoption (Fliegel & Kivlin, 1966a, 1966b; Martin et al., 1988; Sinden & King, 1990).

The following it can be deduced from the Classical Diffusion Theory: innovations that are perceived as advantageous by potential adopters diffuse rapidly; and the higher the

---

[27] Means of Production as defined by the IFCN are: animal purchases, feed (purchase feed, fertiliser, seed, pesticides), machinery (maintenance, depreciation, contractor), fuel, energy, lubricants, water, buildings (maintenance, depreciation), veterinarian and medicine, insemination, insurance and taxes, and other inputs.

advantage of a given innovation over the old ideas, the faster the diffusion of that innovation.

The fact that typical Argentine farms had similar Entrepreneur's Profit to typical New Zealand farms, and the fact that some typical Argentine farms performed as well as the case studies that had adopted a large number of New Zealand innovations, showed that the advantage in Entrepreneur's Profit of the innovations over the traditional practices was not so significant. This could be the main reason why New Zealand innovations have not been widely adopted in Argentina. However, this is not the only reason, other constraints can be related to compatibility issues or other factors proposed by the Diffusion Theory.

Despite the fact that benefits of adoption were not so clearly seen in differences in Entrepreneur's Profit, this study defined a number of benefits from adopting New Zealand innovations that should be considered by Argentine dairy farmers and consultants. Additionally, despite the fact that Entrepreneur's Profit is an important indicator of profitability, some of the farmers mentioned that they were focused in maximizing returns per hectare or per capital invested. Therefore their focus was on ROI, in which increases were associated with the adoption of New Zealand innovations.

10 CONCLUSIONS

10.1 The Research Questions

Why New Zealand principles and practices, despite the fact that they appear to be so beneficial to Argentine farmers, have not been widely adopted in Argentine farms?

Can Argentine dairy farmers benefit from adopting New Zealand dairy farm principles and practices?

10.2 Research Conclusions

Related to Entrepreneur's Profit and Return on Investment

Higher levels of adoption of New Zealand innovations by a group of Argentine dairy farms[28] were associated with higher levels of Return on Investment (ROI). Although, no association was found between level of adoption of New Zealand innovations and the level of Entrepreneur's Profit per kilogram of milk[29] produced.

However ROI is increasingly been considered as a more relevant financial indicator for dairy farmers in New Zealand (Nicola Shadbolt, personal communication) and also some of the Argentine farmers mentioned that they were focused in maximizing returns of their investment.

Related to the Cost Component "Land"

Higher levels of adoption of New Zealand innovations by a group of Argentine dairy farms were associated with reductions in land costs per kilogram of milk produced.

The main advantage of the adoption of New Zealand innovations found in the case study farms (especially "increments of phosphate levels in soils", "utilization of

---

[28] Seven real Argentine dairy farms and three typical Argentine dairy farms (IFCN) were studied for the 2002 and 2003 seasons.
[29] All milk considered is Energy Converted Milk (ECM).

quantitative pasture monitoring", and "utilization of pasture budgets"), was the association between the level of adoption and level of milk production per hectare[30].

Related to the Cost Component "Labour"

Higher levels of adoption of New Zealand innovations by a group of Argentine dairy farms were associated with higher levels of labour costs per kilogram of milk produced, because of higher average wages paid per hour of work, and despite higher levels of labour productivity[31].

The New Zealand principle "less than 15 cows milked per set of teat-cups" was found to be the innovation most closely associated with increases in labour productivity. Other innovations adopted by the Argentine farmers that could be associated with increases in labour productivity were: seasonal calving, rearing of calves in groups, and New Zealand style milking sheds and systems.

Additionally, the adoption of New Zealand innovations was associated with increasing levels of formal education of people working on dairy farms, which was also associated with higher wages paid.

Related to the Cost Component "Capital" (land not included)

A slight association was found between the level of adoption of New Zealand innovations by a group of Argentine dairy farms and the level of capital costs per kilogram of milk produced. Additionally the adoption of New Zealand innovations was associated with reduction in capital productivity[32].

Related to the Cost Component "Means of Production"

A slight association was found between the adoption of New Zealand innovations by a group of Argentine farms and higher costs as means of production[33] per kilogram of milk produced.

---

[30] Land productivity measured in kilograms of milk (ECM) per hectare.
[31] Labour productivity measured in kilograms of milk (ECM) per hour of work, including all people that worked in the dairy enterprise (contract labour not included).
[32] Capital productivity measured in kilograms of milk (ECM) produced per US-$1000 invested on buildings, machinery, livestock, and others.
[33] Means of Production as defined by the IFCN are: animal purchases, feed (purchase feed, fertiliser, seed, pesticides), machinery (maintenance, depreciation, contractor), fuel, energy, lubricants, water,

Related to New Zealand Genetics

Cows of New Zealand genetics were believed to be necessary for the adoption of seasonal calving, by the Argentine case study farmers and dairy sector experts. An association was found between the adoption of New Zealand genetics and higher milk yield per kilogram of live weight. Additionally, farms that adopted New Zealand genetics had lower mortality and replacement rates than those that had not adopted them.

10.3 Assessment of the methodology

The present study had to deal with a general lack of accurate information about the Argentine production sector. No exhaustive survey of the population of Argentine dairy farmers has been done. However, an extensive survey done by Gambuzzi et al. (2003)[34] and the average and typical farms defined by the panel of Argentine dairy production experts of the IFCN were very useful in order to place the case study farms within the population of Argentine dairy farms.

It should be mentioned that at the time when this research was completed no other study on adoption of New Zealand innovations by Argentine farms could be found. Nor were there any reliable information about the following: a list of New Zealand practices, technologies and ideas that would be innovative and useful for Argentine farmers; the proportion, number or identity of the Argentine dairy farmers that were adopting New Zealand ideas; and the New Zealand ideas that had already been adopted by the farmers in contact with the New Zealand dairy farm systems were also unknown.

Related to the Qualitative Data

The case study strategy in general, and the multiple and embedded case study in particular, were found to be appropriate in order to meet the first four objectives of the present study (see section 1.2). The chosen design was useful to define a group of New

---

buildings (maintenance, depreciation), veterinarian and medicine, insemination, insurance and taxes, and other inputs.

[34] Gambuzzi et al. (2003) did a survey of 530 dairy farms on most of the main production areas of Argentina. However the sample was done randomly and therefore it is not known how representative it is. Additionally, not all the production areas of Argentina were surveyed (for example: there were no dairy farms representing the south east of Buenos Aires province).

Zealand innovations that were potentially useful for Argentine dairy farms, to assess the level of adoption of these innovations, and to identify the reasons for their adoption and rejection.

In order meet the fourth objective (assess the impact of the adoption of New Zealand innovations on the performance of the case study farms), it was necessary to add some typical Argentine farms to the case study farms. The typical Argentine farms from the IFCN —which were assumed that had not adopted any New Zealand innovation- were useful for comparison with the case studies -which had adopted different levels of New Zealand innovations. Some associations were found between the level of adoption of the innovations and the level of some physical and financial performances. However a greater number of farms, and possibly data from several seasons, would have been necessary in order to examine the possible relationships between the adoption of New Zealand innovations and some physical and economical variables statistically.

The fifth objective of the present study (assess which have been the main causes of the non-spread of New Zealand innovations in Argentine dairy farms) was partially accomplished by identifying, what was possibly the main cause for non-spread of New Zealand innovations. However, it was not possible to investigate other issues related to the diffusion of the innovations due to time constraints.

Related to the Quantitative Data

The IFCN database provided a valuable framework for the quantitative analysis, based on four costs components (land, labour, capital and means of production). This framework was very helpful in assessing the impact of adopting New Zealand innovations on the performance of the case study farms. This framework, which was designed to contrast farms from different countries, enabled the comparisons between the Argentine and New Zealand typical farms, and also between the typical and real Argentine dairy farms. The IFCN database encloses some innovative physical and financial indicators, as for example: "Costs of Milk Production Only" per kilogram of milk produced; "Non-milk Returns" per kilogram of milk produced; "Entrepreneur's Profit" per kilogram of milk produced; and the ways of calculating the opportunity costs of land, labour and capital. Additionally the computer spreadsheet TIPI-CAL was useful to process the data from the case study farms and to illustrate the results in graphs.

10.4 Further Research

- The impact of adopting each New Zealand innovation on the performance of Argentine dairy farms should be investigated further, for the impact on ROI; on Operating Profit per hectare; on productivities of land, labour and capital; and on production costs.

- Further research should be done in order to study the trade off between level of investment in plant and machinery in Argentine dairy farms (especially the number of set of teat-cups) and total labour productivity. There should be a level of investment for Argentine dairy farms when labour costs savings (per kilogram of milk produced) are greater than the additional capital costs (per kilogram of milk produced).

- Future research should be done in order to assess which is the optimum type of cow for typical Argentine dairy systems. The similarities and differences between the Argentine, the New Zealand and Irish dairy systems, in which New Zealand genetics cows' performance is being investigated, should be explored.

- Further research investigating the benefits of adopting New Zealand innovations by farms with different proportions of grazed pasture in the total annual diet for the herd.

- The main goals of Argentine dairy farmers should be studied. And also which indicators should be taken into account by typical Argentine farmers in order to achieve their dairy farm goals.

- Further research should be done to assess the opportunities and constraints for the diffusion of New Zealand innovations in the future. The Classic Theory of Diffusion could provide a helpful framework in order to accomplish this objective. This might lead to a more general study of advisory and extension services requires in Argentina.

- The utilization of debt in dairy farming in Argentina should be studied; and also the possible returns of Argentine dairy farms and the risks involved at different levels of debt.

11 REFERENCES

AACREA. (1986). Milk production 1. In *Cuaderno de Actualización Técnica CREA.*

ACHA (2000). *Asociación Criadores de Holando Argentino.* Retrieved 7/01, 2005, from http://www.viarural.com.ar/viarural.com.ar/ganaderia/asociaciones/holando/elho landoargentino01.htm

Arnon, I. (1989). *Agricultural research and technology transfer.* London, New York: Elsevier Applied Science.

BCRA. (2005). *Presentación del Programa Monetario 2005 ante el Honorable Senado de la Nación.* Retrieved 15/03, 2005, from http://www.bcra.gov.ar/hm000000.asp

Black, A. W. (2000). Extension theory and practice: a review. *Australian Journal of Experimental Agriculture, 40,* 493-502.

Bouma, G. D. (2000). *The research process* (4th ed.). Melbourne: Oxford University Press.

Campbell, A., & Junor, B. (1992). Land management extension in the '90s: Evolution or emasculation? *Australian Journal of Soil and Water Conservation, 5*(2).

Caplan, N., & Nelson, S. D. (1973). On Being Useful: The Nature and Consequences of Psychological Research on Social Problems. *American Psychologist, 28,* 199-211.

Carberry, P. S., Hochman, Z., McCown, R. L., Dalgliesh, N. P., Foale, M. A., Poulton, P. L., et al. (2002). The FARMSCAPE approach to decision support: Farmers', advisers', researchers', monitoring, simulation, communication and performance evaluation. *Agricultural Systems, 74,* 141-177.

Castignani, H., & Zehnder, R. (2003). *Análisis de Empresas Tamberas de Cambio Rural de la Zona Centro de la Provincia de Santa Fe - Año 2001/2002.* Retrieved 10/2004, 2004, from http://www.inta.gov.ar/rafaela/info/documentos/economia

Chambers, R., & Ghildyal, B. P. (1985). Agricultural-Research for Resource-Poor Farmers - the Farmer-1st-and-Last Model. *Agricultural Administration, 20*(1), 1-30.

Chambers, R., & Jiggins, J. (1987). Agricultural-Research for Resource-Poor Farmers .1. Transfer-of-Technology and Farming Systems Research. *Agricultural Administration and Extension, 27*(1), 35-52.

Chetty, S. (1996). The Case Study Method for Research in Small- and Medium-sizes Firms. *International Small Business Journal, 15,* 73-85.

CIL (2004). *Exportación de Productos Lácteos (detalle mensual)*. Retrieved 18/01, 2004, from http://www.cil.org.ar/

Clark, D. A., Carter, W., Walsh, B., Clarkson, F. H., & Waugh, C. D. (1994). *Effect of winter pasture residuals and grazing off on subsequent milk production and pasture performance*. Paper presented at the New Zealand Grasslands Association Conference.

Dexcel. (2003). *Benchmarking Efficient Farm Systems*.

Eisenhardt, K. M. (1989). Building Theories from Case-Study Research. *Academy of Management Review, 14*(4), 532-550.

Eisenhardt, K. M. (1991). Better Stories and Better Constructs - the Case for Rigor and Comparative Logic. *Academy of Management Review, 16*(3), 620-627.

Euromonitor. (2004a). *Argentina, Country Profile*. Retrieved 25/08, 2004, from http:/www.euromonitor.com/gmid/printdata.asp

Euromonitor. (2004b). *Countries Profiles*. Retrieved 11/06, 2004, from http://www.euromonitor.com/gmid/default.asp

Euromonitor. (2004c). *New Zealand, Country Profile*. Retrieved 14/06, 2004, from http:/www.euromonitor.com/gmid/printdata.asp

Fencepost.com. (2004, 21/07/2004). *Fonterra Payout Lifts 2 Cents To $4.25*. Retrieved 20/01, 2004, from http://www.fencepost.com/dairy/industry/detail.jhtml?ElementID=/dairy/news/r epository/20040721_090438_Fonterra_Payout_Lifts_2_Cents_To_4_25.xml

Fliegel, F. C., & Kivlin, J. E. (1966a). Attributes of Innovations as Factors in Diffusion. *The American Journal of Sociology, 72*(2), 235-248.

Fliegel, F. C., & Kivlin, J. E. (1966b). Farmers' Perceptions of Farm Practice Attributes. *Rural Sociology, 31*(2), 197-206.

FXHistory: Historical currency exchange rates. (2005). Retrieved 08/03, 2005

Gambuzzi, E. L., Zehnder, R., & Chimicz, J. (2003). *Análisis de Sistemas de Producción Lechera*. Retrieved September 17th, 2004, from http://www.inta.gov.ar/rafaela/info/documentos/economia/analisis_sistemas_pro d_lechera.pdf

Garcia, S. C. (1997). Milk Production in Argentina. *Dairyfarming Annual, 49*, 86-91.

Garcia, S. C., & Holmes, C. W. (1999). Effects of time of calving on the productivity of pasture-based dairy systems: A review. *New Zealand Journal of Agricultural Research, 42*, 347-362.

Goldman, K. D. (1994). Perceptions of Innovations as Predictors of Implementation Levels - the Diffusion of a Nationwide Health-Education Campaign. *Health Education Quarterly, 21*(4), 429-444.

Goss, K. F. (1979). Consequences of Diffusion of Innovations. *Rural Sociology, 44,* 754-772.

Gray, D. I. (2001). *The tactical management processes used by pastoral-based dairy farmers: A multiple-case study of experts.* Unpublished thesis presented in partial fulfilment of the requirements for the degree of Doctor of Philosophy in Farm Management at Massey University.

Guardini, E., Labriola, S., & Schaller, A. (1999, July). Cadenas Alimentarias, Productos Lacteos. *Alimentos Argentinos.*

Guerin, L. J., & Guerin, T. F. (1994). Constraints to the Adoption of Innovations in Agricultural-Research and Environmental-Management - a Review. *Australian Journal of Experimental Agriculture, 34*(4), 549-571.

Guerin, T. F. (1999). An Australian perspective on the constraints to the transfer and adoption of innovations in land management. *Environmental Conservation, 26*(4), 289-304.

Gutman, G., Guiguet, E., & Rebolini, J. (2003). *Los Ciclos en el Complejo Lácteo Argentino. Análisis de Políticas Lecheras en Países Seleccionados*: SAGPyA.

Haider, M., & Kreps, G. L. (2004). Forty years of Diffusion of Innovations: Utility and value in public health. *Journal of Health Communication, 9*(Suplement), 3-11.

Harrison, W. R., & Kennedy, L. P. (1997). A Neoclassical Economic and Strategic Management Approach to Evaluating Global Agribusiness Competitiveness. *Competitive Review, 7*(1), 14-27.

Hodgson, J. (1990). *Grazing management: Science into practice*: Longman Scientific & Technical.

Holmes, C. W. (2003a). *For all systems (high and low inputs), costs per kg milksolids must be appropriate for the system's output of kg milksolids per hectare.* Palmerston North, NZ: Massey University.

Holmes, C. W. (2003b). *Low Cost Production of Milk from Grazed Pastures: An outline of dairy production systems in New Zealand.* Palmerston North, N.Z.: Massey University.

Holmes, C. W., Brooks, I. M., Garrick, D. J., MacKenzie, D. D. S., Parkinson, T. J., & Wilson, G. F. (2002). *Milk production from pasture.* Palmerston North, N.Z.: Massey University.

Hopkins, A. (2000). *Grass: Its production and utilization* (3rd ed.). Oxford: Blackwell Science.

Howden, P., & Vanclay, F. (2000). Mythologization of farming styles in Australian broadacre cropping. *Rural Sociology, 65*(2), 293-310.

IFCN. (2002). *Dairy Report. Status and Prospects of Typical Dairy Farms World-Wide.* (No. 1610-434X). Braunschweig, Germany: IFCN (International Farm Comparison Network).

IFCN. (2003). *IFCN Dairy Report. For a Better Understanding of Dairy Farming World-Wide.* (No. 1610-434X): IFCN (International Farm Comparison Network).

IFCN. (2004). *IFCN Dairy Report. For a Better Understanding of Dairy Farming World-Wide.*: IFCN (International Farm Comparison Network).

IFCN Dairy Report. For a Better Understanding of Dairy Farming World-Wide. (No. 1610-434X)(2003). IFCN (International Farm Comparison Network).

INDEC. (2003). *Encuesta Industrial Anual 2000 y 2001.* Retrieved 27/03, 2005, from http://www.alimentosargentinos.gov.ar/estadisticas/Encuesta_Industrial_Anual_2000-2001.xls

INDEC. (2005). *Sistema de Indices de Precios al por Mayor (SIPM).* Retrieved 08/03, 2005

Isermeyer, K., Deblitz, C., Hemme, T., & Plessmann, F. (2004, March 2004). *20 Questions and Answers about IFCN.* Retrieved 14/09/2004, 2004, from http://www.ifcnnetwork.org/Downloads/e-20Qmar04.pdf

Kaplan, A. W. (1999). From passive to active about solar electricity: Innovation decision process and photovoltaic interest generation. *Technovation, 19*(8), 467-481.

Kincaid, L. D. (2004). From Innovation to Social Norm: Bounded Normative Influence. *Journal of Health Communication, 9*(Suplement), 37-57.

La Opinión, R.-A. (2004). *La exportación de lácteos sería récord.* Retrieved 15/03, 2005, from http://www.laopinion-rafaela.com.ar/opinion/2004/10/21/p4a2109.htm

Lambourne, A., & Betterridge, K. (2004). *What is more Important - the Pasture or the Cow?* Paper presented at the Dairy 3, Rotorua, New Zealand.

LIC. (2000/01). *Dairy statistics.* Hamilton, N.Z: Livestock Improvement Corporation.

LIC. (2001). *Company Profile / Strengths of Livestock Improvement New Zealand Genetics Cattle.* Retrieved 7/01, 2005, from http://www.newzealandgenetics.com/main.cfm?id=1

LIC. (2001/02). *Dairy statistics.* Hamilton, N.Z: Livestock Improvement Corporation.

LIC. (2002/03). *Dairy statistics.* Hamilton, N.Z.: Livestock Improvement Corporation.

LIC. (2003/04). *Dairy statistics.* Hamilton, N.Z.: Livestock Improvement Corporation.

Lievrouw, L. A., & Pope, J. T. (1994). Contemporary-Art as Aesthetic Innovation - Applying the Diffusion-Model in the Art World. *Knowledge-Creation Diffusion Utilization, 15*(4), 373-395.

Lockie, S., Mead, A., Vanclay, F., & Butler, B. (1995). Factors encouraging the adoption of more sustainable crop rotations in south-east Australia: profit, sustainability, risk and stability. *Journal of Sustainable Agriculture, 6*(1).

Luna, F. (1994/95). *Historia integral de la Argentina*. Buenos Aires: Planeta.

Lynch, T., Gregor, S., & Midmore, D. (2000). Intelligent support systems in agriculture: how can we do better? *Australian Journal of Experimental Agriculture, 40*(4), 609-620.

MacDonald, K. A., Penno, J. K., Nicholas, P. K., Lile, J. A., Coulter, M., & Lancaster, J. A. S. (2001). *Farm Systems - Impact of Stocking Rate on Dairy Farm Efficiency*. Paper presented at the New Zealand Grassland Association.

MAF. (2004). *Dairy Monitoring - Sector Overview*. Retrieved 20/01, 2004, from http://www.maf.govt.nz/mafnet/rural-nz/statistics-and-forecasts/farm-monitoring/2004/dairy/dairy-03.htm#TopOfPage

Martin, R. J., McMillan, M. G., & Cook, J. B. (1988). Survey of Farm-Management Practices of the Northern Wheat Belt of New-South-Wales. *Australian Journal of Experimental Agriculture, 28*(4), 499-509.

Mayne, S. (1998). *Selecting the correct dairy cow for grazing systems*. Paper presented at the Ruakura Farmers' Conference.

McCall, D. G., & Sheath, G. W. (1993, 8-21 February). *Development of Intensive Grassland Systems: From Science to Practice*. Paper presented at the XVII International Grassland Congress, New Zealand: Palmerston North, Hamilton and Lincoln; Australia: Rockhampton.

Mensch, B. S., Bagah, D., Clark, W. H., & Binka, F. (1999). The changing nature of adolescence in the Kassena-Nankana district of northern Ghana. *Studies in Family Planning, 30*(2), 95-111.

Meyer, G. (2004). Diffusion Methodology: Time to innovate? *Journal of Health Communication, 9*(Suplement), 59-69.

Mintzberg, H. (1973). *The nature of managerial work*. Englewood Cliffs, N.J.: Prentice-Hall.

Mitchell, N. (2002). *The Dairy Industry in New Zealand - New Structure and Future*. Paper presented at the Implication of policy changes for the world dairy Industry. Proceedings of the Policy, Economics & Marketing Conference, IDF World Summit 2001 NZ.

Molinuevo, H. A. (2001). *Ajuste entre el potencial genético-productivo y el sistema de producción lechera*. Retrieved 10/09/2004, 2004, from http://www.inta.gov.ar/balcarce/info/documentos/ganaderia/bovinos/genetica/genetica.htm

Moore, G. C., & Benbasat, I. (1991). Development of an Instrument to Measure the Perceptions of Adopting an Information Technology Innovation., *Information*

Systems Research (Vol. 2, pp. 192): INFORMS: Institute for Operations Research.

Murray, P. (2000). Evaluating participatory extension programs: Challenges and problems. *Australian Journal of Experimental Agriculture, 40*(4), 519-526.

Murthy, B. K., Dudhani, C. M., Jayaramaiah, K. M., Veerabhadraiah, V., & Sethu Rao, M. K. (1973). A scale to measure the differential perceptions of adopters and non-adopters. *Mysore Journal of Agricultural Science, 7*(2).

NZ-Biosecurity. (1997). *Code of Recommendations and Minimum Standards for the Welfare of Bobby Calves.* Retrieved 13/03, 2005, from http://www.biosecurity.govt.nz/animal-welfare/codes/bobby-calves/

Ostrowski, B., & Deblitz, C. (2001). *La competitividad en producción lechera de los países de Chile, Argentina, Uruguay y Brasil.*

Pande, T. N., Valentine, I., Betteridge, K., MacKay, A., & Horne, D. (2000). *Pasture damage and regrowth from cattle treating.* Paper presented at the New Zealand Grassland Association Conference.

Penno, J. (1999). *Stocking Rate for Optimum Profit.* Paper presented at the South Island Dairy Event.

Philliber, S. G., Schwab, G., & Sloss, S. (1980). *Social research.* Itasca: F. E. Peacock Publishers.

Ritson, C. (1977). *Agricultural economics: principles and policy.* London: Crosby Lockwood Staples.

Rogers, E. M. (2003). *Diffusion of innovations* (5th ed.). New York: Free Press.

Rogers, E. M. (2004). A prospective and retrospective look at the diffusion model. *Journal of Health Communication, 9*(Suplement), 13-19.

Rogers, E. M., & Kincaid, L. D. (1981). *Communication networks: toward a new paradigm for research.* New York, London: Free Press; Collier Macmillan.

Rogers, E. M., & Shoemaker, F. F. (1971). *Communication of innovations; a cross-cultural approach* (2nd ed.). New York: Free Press.

Röling, N. G. (1988). *Extension science: Information systems in agricultural development.* Cambridge; New York: Cambridge University Press.

Röling, N. G., Ascroft, J., & Wa Chege, F. (1981). The Diffusion of Innovations and The Issue of Equity in Rural Development. In B. R. Crouch & S. Chamala (Eds.), *Extension Education and Rural Development* (Vol. 1, pp. 225-236). Chichester, New York, Brisbane, Toronto.

Rougoor, C. W., Trip, G., Huirne, R. B. M., & Renkema, J. A. (1998). To define and study farmers' management capacity: theory and use in agricultural economics. *Agricultural Economics, 18*(3), 261-272.

Rowley, J. (2002). Using case studies in research. *Management Research News, 25*(1), 16-27.

SAGPyA. (2003). *Informe de coyuntura de Lácteos No.20.* Retrieved 20/06, 2003, from http://www.alimentosargentinos.gov.ar/lacteos/default.asp

SAGPyA. (2004a). *Producción Argentina de Leche.* Retrieved 27/08, 2004, from http://www.alimentosargentinos.gov.ar/0-3/lacteos/02_Nacional/serie/Prod_Anual.htm

SAGPyA. (2004b). *Informe de coyuntura de Lácteos No.26.* Retrieved 30/08, 2004, from http://www.alimentosargentinos.gov.ar/lacteos/default.asp

Sinden, J. A., & King, D. A. (1990). *Articles and Notes:* Adoption of Soil Conservation Measures in Manilla Shire, New South Wales. *Review of Marketing and Agricultural Economics, 58*(2-3), 179-192.

Statistics-NZ. (2000). *Quick facts Industries: Dairy products.* Retrieved 30/06, 2004, from http://www.stats.govt.nz/domino/external/Web/nzstories.nsf/092edeb76ed5aa6b cc256afe0081d84e/b88ff0f2aa375339cc256b1f0000ebc1?OpenDocument

Statistics-NZ. (2003). *New Zealand External Trade Statistics.* Retrieved 30/07, 2004, from http://www.stats.govt.nz/domino/external/web/prod_serv.nsf/092edeb76ed5aa6b cc256afe0081d84e/ea7ca8e004467b49cc256e43007c20b7/%24FILE/NZETS-Dec03.pdf

Thomson, N. A., & al, e. (1993). *Winter grazing management; effects on pastures.* Paper presented at the Dairyfarming Annual, Massey University.

Thomson, N. A., Roberts, A. H. C., McCallum, D. A., Judd, T. G., & Johnson, R. J. (1993). *How Much Phosphate Fertiliser is Enough?* Paper presented at the Ruakura Farmers Conference.

Tully, J. (1981). Changing Practices: A Case-study. In B. R. Crouch & S. Chamala (Eds.), *Extension Education and Rural Development* (Vol. 2, pp. 79-86). Chichester, New York, Brisbane, Toronto.

USDA. (2004a). *Dairy Exports, Selected Countries (Butter, Non-fat dry milk and Cheese).* Retrieved 20/08, 2004, from http://www.fas.usda.gov/psd/intro.asp?circ_id=11

USDA. (2004b, 19/07/2004). *USDA.* Retrieved 20/08/2004, from http://www.fas.usda.gov/psd/intro.asp?circ_id=11

Valente, T. W., & Rogers, E. M. (1995). The Origins and Development of the Diffusion of Innovations Paradigm as an Example of Scientific Growth. *Science Communication, 16*(3), 242-273.

Vanclay, F. (2004). Social principles for agricultural extension to assist in the promotion of natural resource management. *Australian Journal of Experimental Agriculture, 44*(3).

Veerabhadraiah, V., Sethu Rao, M. K., & Dwarakinath, R. (1973a). Identification of opinion leaders in a rural community and their characteristics. *Mysore Journal of Agricultural Sciences, 7*(2).

Veerabhadraiah, V., Sethu Rao, M. K., & Dwarakinath, R. (1973b). Influence of referents on adoption of farm practices in two villages of Dharwar community development block. *Mysore Journal of Agricultural Sciences, 7*(2).

Warner, K. (1974). The need for some innovative concepts of innovation: an examination of research on the diffusion of innovations. *Policy Sciences, 5*(4).

Yin, R. K. (2002). *Case study research: design and methods* (3rd ed.). Thousand Oaks, California: Sage Publications.

Zaltman, G. (2003). *How customers think: Essential insights into the mind of the market.* New York, London: McGraw-Hill.

www.ingramcontent.com/pod-product-compliance
Lightning Source LLC
Chambersburg PA
CBHW060356220326
41598CB00023B/2941